Sports Materials

Sports Materials

Special Issue Editors

Thomas Allen
Leon Foster
Martin Strangwood
James Webster

MDPI • Basel • Beijing • Wuhan • Barcelona • Belgrade

Special Issue Editors

Thomas Allen
Manchester Metropolitan University
UK

Leon Foster
Sheffield Hallam University
UK

Martin Strangwood
The University of Birmingham
UK

James Webster
Under Armour Inc.
USA

Editorial Office
MDPI
St. Alban-Anlage 66
4052 Basel, Switzerland

This is a reprint of articles from the Special Issue published online in the open access journal *Applied Sciences* (ISSN 2076-3417) from 2017 to 2019 (available at: https://www.mdpi.com/journal/applsci/special_issues/Sports_Materials)

For citation purposes, cite each article independently as indicated on the article page online and as indicated below:

LastName, A.A.; LastName, B.B.; LastName, C.C. Article Title. *Journal Name* **Year**, *Article Number*, Page Range.

ISBN 978-3-03928-162-6 (Pbk)
ISBN 978-3-03928-163-3 (PDF)

Contents

About the Special Issue Editors vii

Thomas Allen, Leon Foster, Martin Strangwood and James Webster
Sports Materials Special Issue Editorial
Reprinted from: *Appl. Sci.* **2019**, *9*, 5272, doi:10.3390/app9245272 . . . . **1**

Tom W. Corke, Nils F. Betzler, Eric S. Wallace, Martin Strangwood and Steve R. Otto
Implications of Rigid Gripping Constraints on Clubhead Dynamics in Steel Golf Shafts
Reprinted from: *Appl. Sci.* **2018**, *8*, 422, doi:10.3390/app8030422 . . . . **4**

Ben Lane, Paul Sherratt, Xiao Hu and Andy Harland
Measurement of Strain and Strain Rate during the Impact of Tennis Ball Cores
Reprinted from: *Appl. Sci.* **2018**, *8*, 371, doi:10.3390/app8030371 . . . . **12**

Renaud G. Rinaldi, Lionel Manin, Sébastien Moineau and Nicolas Havard
Table Tennis Ball Impacting Racket Polymeric Coatings: Experiments and Modeling of Key
Performance Metrics
Reprinted from: *Appl. Sci.* **2019**, *9*, 158, doi:10.3390/app9010158 . . . . **24**

Joshua Fortin-Smith, James Sherwood, Patrick Drane and David Kretschmann
Characterization of Maple and Ash Material Properties for the Finite Element Modeling of
Wood Baseball Bats
Reprinted from: *Appl. Sci.* **2018**, *8*, 2256, doi:10.3390/app8112256 . . . . **40**

**Joshua Fortin-Smith, James Sherwood, Patrick Drane, Eric Ruggiero, Blake Campshure and
David Kretschmann**
A Finite Element Investigation into the Effect of Slope of Grain on Wood Baseball Bat Durability
Reprinted from: *Appl. Sci.* **2019**, *9*, 3733, doi:10.3390/app9183733 . . . . **56**

David Cole, Steph Forrester and Paul Fleming
Mechanical Characterisation and Modelling of Elastomeric Shockpads
Reprinted from: *Appl. Sci.* **2018**, *8*, 501, doi:10.3390/app8040501 . . . . **71**

**Francesco Penta, Giuseppe Amodeo, Antonio Gloria, Massimo Martorelli,
Stephan Odenwald and Antonio Lanzotti**
Low-Velocity Impacts on a Polymeric Foam for the Passive Safety Improvement of Sports
Fields: Meshless Approach and Experimental Validation
Reprinted from: *Appl. Sci.* **2018**, *8*, 1174, doi:10.3390/app8071174 . . . . **84**

**Leon Foster, Prashanth Peketi, Thomas Allen, Terry Senior, Olly Duncan and
Andrew Alderson**
Application of Auxetic Foam in Sports Helmets
Reprinted from: *Appl. Sci.* **2018**, *8*, 354, doi:10.3390/app8030354 . . . . **97**

**Olly Duncan, Todd Shepherd, Charlotte Moroney, Leon Foster, Praburaj D. Venkatraman,
Keith Winwood, Tom Allen and Andrew Alderson**
Review of Auxetic Materials for Sports Applications: Expanding Options in Comfort
and Protection
Reprinted from: *Appl. Sci.* **2018**, *8*, 941, doi:10.3390/app8060941 . . . . **109**

Jia-Horng Lin, Chih-Hung He, Yu-Tien Huang and Ching-Wen Lou
Functional Elastic Knits Made of Bamboo Charcoal and Quick-Dry Yarns: Manufacturing Techniques and Property Evaluations
Reprinted from: *Appl. Sci.* **2017**, *7*, 1287, doi:10.3390/app7121287 . . . . **142**

About the Special Issue Editors

Thomas Allen (Dr.) is a senior lecturer in Mechanical Engineering at Manchester Metropolitan University, where he specializes in Sports Engineering. He is interested in the effect of engineering and technology on sport, in terms of performance, participation and injury risk. His research involves finite element modelling and the mechanical testing of sports equipment, as well as the development and characterization of materials for sporting applications. Tom is an active member of the International Sports Engineering Association (ISEA); he was one of the organizers of their 10th conference, and is Editor in Chief of their journal Sports Engineering. He is also a fellow of the IMechE, and serves on the board of the International Society for Snowsport Safety (ISSS).

Leon Foster (Dr.) is a researcher at the Centre for Sports Engineering Research, specialising in performance analysis systems. Leon has developed software and hardware for performance analysis tools for elite athletes. He has also become involved in the development of these technologies for use by the general public through a jointly funded European Union Interreg project called ProFit. Leon has a keen interest in testing sports equipment and has developed testing protocols to evaluate the safety of protective sporting equipment within the Centre for Sports Engineering Research laboratory. In addition to his research, Leon lecturers within the MSc Sports Engineering course, and supervises several Ph.D. students.

Martin Strangwood (Dr.) gained his BA in Natural Sciences from Cambridge University in 1984 and followed that with a Ph.D. in 'Prediction and Assessment of Weldmetal Microstructures' in the Department of Materials Science and Metallurgy at Cambridge in 1987. He spent 3 years at AEA Technology in the Fracture and Materials Evaluation Department of the Harwell Laboratory. During that time he worked on a number of projects, ranging from the thermal degradation of stainless steels to the development of metal matrix composites and intermetallics. In 1990, he joined the School of Metallurgy and Materials at the University of Birmingham, where he has established the Phase Transformations and Microstructural Modelling and Sports Materials Research Groups.

His research interests lie in the quantification of structures and properties for a range of materials, relating these to processing conditions, composition and eventual properties. The physical relationships established are then used with thermodynamic and kinetic models to optimise processing and composition combinations. Much work has been carried out on various grades of steel, although aluminium-, titanium-, nickel- and copper-based alloys also feature. In the sports engineering field this has been applied to golf clubs (in collaboration with the R & A) and bicycle frames (working with Reynolds). Other sports examples have included the use of wood, elastomers and composites in cricket, hockey, golf and cycling. He was editor of Sports Engineering for 6 years.

He is now a senior lecturer at Birmingham and has published over 200 papers, along with supervising 70 Ph.D. and Masters students.

James Webster (Dr.) (Senior Manager) is an innovation leader at Under Armour, where he specializes in material development, spanning both footwear and apparel, with a particular interest in polymer development. His work incorporates a holistic approach to innovation, having a background in human performance and mechanical engineering, which he incorporates into his day to day work in developing novel technologies for the sports market. James is an active member of the international sports engineering associations executive committee and is also on the editorial board for the Journal of Sports Engineering and Technology, where he has been a guest editor on multiple special editions.

Editorial

Sports Materials Special Issue Editorial

Thomas Allen [1,*]**, Leon Foster** [2]**, Martin Strangwood** [3] **and James Webster** [4]

[1] Department of Engineering, Manchester Metropolitan University, Manchester M15 6BH, UK
[2] Centre for Sports Engineering Research, Sheffield Hallam University, Sheffield S10 2BP, UK;
    l.i.foster@shu.ac.uk
[3] School of Metallurgy and Materials, The University of Birmingham, Edgbaston Birmingham B15 2TT, UK;
    m.strangwood@bham.ac.uk
[4] Under Armour Inc, Baltimore 21230, UK; jwebster@underarmour.com
* Correspondence: t.allen@mmu.ac.uk

Received: 26 November 2019; Accepted: 28 November 2019; Published: 4 December 2019

Materials are key to the world of sport. Advances in materials have enhanced equipment and clothing to allow athletes to perform better and set new records, while improving safety and making sport and exercise more accessible, comfortable and enjoyable. Tennis saw dramatic changes in the 1980s, as sports engineers discarded wood and adopted fibre-polymer composites to produce stiffer rackets with larger heads [1]. Fibre-polymer composites are also used to create stiff plates within the midsole of some distance running shoes, allowing athletes to run more efficiently [2,3]. In snowsports, soft-shell back protectors made from shear thickening foam are more comfortable and offer better protection from repeat impacts than their traditional hard-shell counterparts [4]. The work behind such advances in materials and equipment should ideally be communicated in peer reviewed articles, so claims can be verified and improvements realised, and appropriately implemented and regulated, to ensure sport remains fair, safe and enjoyable for all.

This Special Issue on "Sports Materials" brings together 10 articles covering equipment testing, modelling and material development. As an established tool for designing and testing sports equipment and assessing materials [5], finite element modelling is clearly present in this issue. The global appeal of sports engineering is reflected in contributions from China, Taiwan, UK, Italy, Germany, France, Sweden and the USA.

The issue opens with an article from Corke and colleagues [6] on the effect of golf club handle gripping conditions on the ball-clubhead interaction. By measuring strain propagation along the shaft, the time for impact-induced vibrations to travel from the clubhead to the handle and back was shown to exceed the duration of ball-clubhead contact. Therefore, the way in which the club handle is gripped should not affect how the ball leaves the clubhead, meaning golf robots are comparable to human testers. Moving from hand-held equipment to ball dynamics, Lane and colleagues [7] measured strains in tennis ball cores during impacts with a rigid surface, using three-dimensional digital image correlation. Knowledge of the strains experienced by sports equipment during use can inform material characterisation and finite element modelling strategies, with implications for product development and selection.

The first article in the issue utilising finite element modeling is from Rinaldi and colleagues [8], and covers the oblique impact of a table tennis ball on the polymeric layers that are applied to the faces of wooden bats. Sports engineering is not always about product development and the latest innovations, with regulators often more concerned with maintaining tradition and ensuring fair and safe play. Fortin-Smith and coworkers [9,10] present two articles on finite element models for investigating wooden baseball bat durability and failure mechanisms. They are interested in preventing potentially hazardous multi-piece failures, where a fragment of a broken bat can become a dangerous high-speed projectile. In the first article [9] they characterised the wood used in bats, developed the models and compared them to experiments, while in the second article [10] they used the models to investigate the effect of the slope of grain of the wood on bat failure. Cole et al. [11] developed finite element modelling techniques for predicting the response of artificial turf shock pads to vertical impact loading, with implications for surface design and regulation, as well as player-surface interactions.

Three articles in the issue cover materials for safety devices. The first of these articles was authored by Penta and coworkers [12], and showcases a meshless approach for modelling the impact response of polymeric foam crash mats, with implications for the design of foam protective devices. The meshless approach allowed the crash mat model to function under high strain deformation, where elements within a mesh are prone to distortion. Foster and colleagues [13] explored auxetic open cell foam as a replacement for the conventional open cell foam typically used as a comfort layer in sports helmets, showing potential for these alternative materials to reduce hazardous impact induced accelerations. Duncan and colleagues [14] present a comprehensive review of auxetic materials for sports applications, covering foams, additively manufactured structures and textiles. While these auxetic materials are appearing in sports products like helmets and athletic shoes, further work is required in their testing, modeling and manufacturing before their potential can be fully realised. The final article in the issue is from Lin et al., [15] and is concerned with the development of functional composite yarns for sportswear applications.

This Special Issue represents the current state-of-the-art in sport materials research. As athletes strive for ever increasing gains in performance and safety, and with consumers demanding higher levels of comfort, customisation and sustainability, we expect sporting goods brands to bring further materials driven innovations to their products. We see finite element modelling as a useful tool in the research and development of sports equipment and clothing utilising novel materials. We urge sports engineers to work closely with sports scientists to investigate equipment-athlete interactions [16,17], to ensure the improvements they envisage become reality.

Acknowledgments: The authors would like to thank Oliver Duncan for providing feedback on the editorial draft.

Conflicts of Interest: The authors declare no conflict of interest.

References

1. Taraborrelli, L.; Grant, R.; Sullivan, M.; Choppin, S.; Spurr, J.; Haake, S.; Allen, T. Materials Have Driven the Historical Development of the Tennis Racket. *Appl. Sci.* **2019**, *9*, 4352. [CrossRef]
2. Hoogkamer, W.; Kipp, S.; Frank, J.H.; Farina, E.M.; Luo, G.; Kram, R. A comparison of the energetic cost of running in marathon racing shoes. *Sports Med.* **2018**, *48*, 1009–1019. [CrossRef] [PubMed]
3. Hunter, I.; McLeod, A.; Valentine, D.; Low, T.; Ward, J.; Hager, R. Running economy, mechanics, and marathon racing shoes. *J. Sports Sci.* **2019**, *37*, 2367–2373. [CrossRef] [PubMed]
4. Signetti, S.; Nicotra, M.; Colonna, M.; Pugno, N.M. Modeling and simulation of the impact behavior of soft polymeric-foam-based back protectors for winter sports. *J. Sci. Med. Sport.* **2019**, *22*, S65–S70. [CrossRef] [PubMed]
5. Choppin, S.; Allen, T. Special issue on predictive modelling in sport. *Proc. Inst. Mech. Eng. Part P J. Sports Eng. Technol.* **2012**, *226*, 75–76. [CrossRef]
6. Corke, T.W.; Betzler, N.F.; Wallace, E.S.; Strangwood, M.; Otto, S.R. Implications of Rigid Gripping Constraints on Clubhead Dynamics in Steel Golf Shafts. *Appl. Sci.* **2018**, *8*, 422. [CrossRef]
7. Lane, B.; Sherratt, P.; Hu, X.; Harland, A. Measurement of Strain and Strain Rate during the Impact of Tennis Ball Cores. *Appl. Sci.* **2018**, *8*, 371. [CrossRef]
8. Rinaldi, R.G.; Manin, L.; Moineau, S.; Havard, N. Table Tennis Ball Impacting Racket Polymeric Coatings: Experiments and Modeling of Key Performance Metrics. *Appl. Sci.* **2019**, *9*, 158. [CrossRef]
9. Fortin-Smith, J.; Sherwood, J.; Drane, P.; Kretschmann, D. Characterization of Maple and Ash Material Properties for the Finite Element Modeling of Wood Baseball Bats. *Appl. Sci.* **2018**, *8*, 2256. [CrossRef]
10. Fortin-Smith, J.; Sherwood, J.; Drane, P.; Ruggiero, E.; Campshure, B.; Kretschmann, D. A Finite Element Investigation into the Effect of Slope of Grain on Wood Baseball Bat Durability. *Appl. Sci.* **2019**, *9*, 3733. [CrossRef]

11. Cole, D.; Forrester, S.; Fleming, P. Mechanical Characterisation and Modelling of Elastomeric Shockpads. *Appl. Sci.* **2018**, *8*, 501. [CrossRef]
12. Penta, F.; Amodeo, G.; Gloria, A.; Martorelli, M.; Odenwald, S.; Lanzotti, A. Low-Velocity Impacts on a Polymeric Foam for the Passive Safety Improvement of Sports Fields: Meshless Approach and Experimental Validation. *Appl. Sci.* **2018**, *8*, 1174. [CrossRef]
13. Foster, L.; Peketi, P.; Allen, T.; Senior, T.; Duncan, O.; Alderson, A. Application of Auxetic Foam in Sports Helmets. *Appl. Sci.* **2018**, *8*, 354. [CrossRef]
14. Duncan, O.; Shepherd, T.; Moroney, C.; Foster, L.; Venkatraman, P.D.; Winwood, K.; Allen, T.; Alderson, A. Review of Auxetic Materials for Sports Applications: Expanding Options in Comfort and Protection. *Appl. Sci.* **2018**, *8*, 941. [CrossRef]
15. Lin, J.-H.; He, C.-H.; Huang, Y.-T.; Lou, C.-W. Functional Elastic Knits Made of Bamboo Charcoal and Quick-Dry Yarns: Manufacturing Techniques and Property Evaluations. *Appl. Sci.* **2017**, *7*, 1287. [CrossRef]
16. Stefanyshyn, D.J.; Wannop, J.W. Biomechanics research and sport equipment development. *Sports Eng.* **2015**, *18*, 191–202. [CrossRef]
17. Allen, T.; Choppin, S.; Knudson, D. A review of tennis racket performance parameters. *Sports Eng.* **2016**, *19*, 1–11. [CrossRef]

 *applied sciences*

Article

Implications of Rigid Gripping Constraints on Clubhead Dynamics in Steel Golf Shafts

Tom W. Corke [1,2], Nils F. Betzler [2,3], Eric S. Wallace [1], Martin Strangwood [4] and Steve R. Otto [2,*]

1   Sport and Exercise Sciences Research Institute, Ulster University, Newtownabbey BT37 0QB, UK; tomcorke14@gmail.com (T.W.C.); es.wallace@ulster.ac.uk (E.S.W.)
2   R&A Rules Ltd., St. Andrews KY16 9JD, UK; nils.betzler@qualisys.com
3   Qualisys AB, Kvarnbergsgatan 2, 411 05 Gothenburg, Sweden
4   School of Metallurgy & Materials, The University of Birmingham, Edgbaston, Birmingham B15 2TT, UK; m.strangwood@bham.ac.uk
*   Correspondence: steveotto@randa.org; Tel.: +44-133-446-000

Received: 31 January 2018; Accepted: 2 March 2018; Published: 12 March 2018

Featured Application: The strain propagation findings presented in this paper justify the continued use of golf robots in studies investigating steel-shafted clubhead dynamics at ball impact, given that the gripping mechanism has a negligible effect on the collision dynamics.

Abstract: Research and equipment testing with golf robots offers much greater control and manipulation of experimental variables compared to tests using human golfers. However, whilst it is acknowledged that the club gripping mechanism of a robot is dissimilar to that of a human, there appears to be no scientific findings on the effects of these gripping differences on the clubhead at ball impact. Theoretical and experimental strain propagation rates from the clubhead to the grip and back to the clubhead were determined during robot testing with a 9-iron to determine if this time interval was sufficiently short to permit the gripping mechanism to have an effect on the clubhead during impact. Longitudinal strain appears to propagate the most quickly, but such deflections are likely to be small and therefore of little meaningful consequence. Shaft bending was not a primary concern as modes of large enough amplitude appear to propagate too slowly to be relevant. Torsional strain propagates at a rate which suggests that constraints at the grip end of a golf club could potentially influence impact dynamics for steel shafted irons; however, this effect seems unlikely to be significant, a likelihood that decreases further for longer irons. As such, it is considered reasonable to treat the influence of a robot's gripping mechanism on clubhead dynamics at impact as negligible, and therefore comparisons between robot and human data in this regard are valid.

Keywords: strain propagation; torsion; golf; shaft; clubhead; robot; cannon

---

1. Introduction

Golf robots are an integral part of the process of testing golf clubs as they overcome many of the limitations imposed by player tests. Despite much research into their operation and application, little effort has been dedicated to following up on recommendations to evaluate the effect of the robot gripping mechanism [1–3]. The gripping mechanism commonly adopts the form of a rigid 'clamp', which has been shown to impose very different constraints on a club's dynamic response when compared to both human hand-held and freely-suspended conditions [4]. Given the brevity of a typical impact between club and ball in golf, the influence of the shaft on the dynamics of the clubhead during impact has long been considered negligible [5–7], however there is little empirical evidence to support this assumption. Although the rigid gripping mechanism will undoubtedly have some effect on the dynamics of the club when considering a robot swing in its entirety, the dynamic response of the shaft during the contact period between clubhead and ball is considered to be more pertinent. Whilst some research has been dedicated to achieving a robot swing that is a closer representation to that of golfers' swings [8,9], the period during which clubhead and ball are in contact has not specifically been investigated during robot testing. Measurement of contact time in golf has been well documented (typically around 0.5 ms at realistic inbound velocities [10–12]), however rates at which various modes of vibration travel from the clubhead to the grip are not as widely reported. Experimental results for steel shafts suggest that this duration is around 1 ms [13], although other studies suggest that this would be an order of magnitude smaller for carbon fibre composite shafts [14]. Studies modelling golf club shafts have reported durations of 1–2 ms before any effect of impact is 'felt' by the golfer [15,16], although what was meant by 'felt' with respect to measurement times was not explicitly stated. It is difficult to make judgements as to whether strain propagates at a great enough speed along a golf shaft to influence impact dynamics based on the reported evidence; the majority of studies suggest it does not [13,15,16], although this is likely to depend largely on material properties.

Thus, the purpose of the present study was to evaluate the potential for the rigid clamping at the grip of a golf club to influence clubhead dynamics during ball impact through the measurement of shaft strain propagation rates. It was hypothesised that the gripping mechanism would not exert greater influence on the clubhead during impact than that associated with the hands of a golfer. The hypothesis was tested by measuring the time taken for vibration to travel from the clubhead to the gripping mechanism and subsequently return to the clubhead.

2. Materials and Methods

Shaft strain propagation rates immediately following ball impact on the club face were theoretically determined and experimentally measured, as outlined below.

2.1. Theoretical Propagation Speeds

Different types of clubhead deflection will induce different types of shaft strain, which travel at different speeds. The consequences and propagation speeds of longitudinal, torsional and bending strains are discussed below. The associated theoretical calculations presented are based upon previous work [17], unless otherwise stated. Material properties have been estimated using MatWeb online databases.

2.1.1. Longitudinal Strain

Movement of the clubhead parallel to the long axis of the shaft results in strain along the shaft's longitudinal axis (either tension or compression). A shaft is stiffest in this direction, and as a result, any deflections are likely to be of small magnitude and limited consequence. Longitudinal waves are non-dispersive, and as such, their speed does not depend on their frequency. This speed (c_L) can be calculated using Equation (1),

$$c_L = \sqrt{\frac{E}{\rho}}, \tag{1}$$

where E is the Young's Modulus of the material, and ρ is the material's mass density, which for steel, are approximately 200 GPa and 7850 kg·m^{-3} respectively, resulting in an estimated propagation speed of longitudinal waves in steel of 5048 m·s^{-1}.

2.1.2. Torsional Strain

Rotation of the clubhead about the shaft's long axis results in torsional strain in the shaft. This is arguably the motion of greatest concern, as the restriction of this form of clubhead rotation during impact would reduce the rate at which the club face opens or closes during an off-centre strike (i.e., towards the toe or heel, relative to the clubhead's centre of gravity location), and potentially the penalty associated with it. The speed at which torsional waves propagate (c_T) is calculated in accordance with Equation (2),

$$c_T = \sqrt{\frac{G}{\rho}}\,,$$ (2)

where the shear modulus of the material (G) is used instead of the Young's Modulus (E). A typical shear modulus of steel is 79 GPa, therefore the propagation speed of torsional waves in this material is approximately 3172 m·s^{-1}. Given the relationship between Young's and shear moduli, shown in Equation (3),

$$E = 2G(1 + v),$$ (3)

where v is Poisson's ratio, the shear modulus will always be smaller than the Young's modulus of the same material. Therefore c_T is always smaller than c_L, meaning that the speed of torsional waves will be lower than that of longitudinal waves. Similar to longitudinal waves, torsional waves are non-dispersive.

2.1.3. Bending Strain

The strain created by bending waves is more complex than torsional or longitudinal strain, and is made up of a combination of compressive, tensile and shear strains. Unlike longitudinal and torsional waves, the phase speed of bending waves (c_B) does depend on wave frequency (f) and can be expressed as a proportion of c_L (see Equation (1)) as shown in Equation (4),

$$c_B = \sqrt{2\pi f c_L}\,.$$ (4)

Frequency testing of the shaft used in the experimental stage of this investigation was used to generate an approximation of the frequency of the first bending mode ($f_1 \approx 5.4$ Hz) under a fixed-free condition. Equation (5) [18] can be used to estimate the frequency of the nth bending mode (f_n),

$$f_n \approx 2.81 \left(n - \frac{1}{2} \right)^2 f_1\,,$$ (5)

for which f_2, f_3 and f_4, were calculated as 34.1, 94.8 and 185.9 Hz respectively. Equation (4) was then used to calculate the corresponding wave speeds ($f_1 \approx 414$ m·s^{-1}, $f_2 \approx 1040$ m·s^{-1}, $f_3 \approx 1734$ m·s^{-1} and $f_4 \approx 2428$ m·s^{-1}). Given that the reduction in energy content with increasing n assumes a Gaussian profile, such that energy $\propto n^2$, higher order modes were not of concern as their potential effect was considered negligible.

2.2. Experimental Setup

A 9-iron was chosen as the test club as it represents the shortest iron club of interest (excluding wedges) and therefore the shortest distance for the waves to travel. A steel-shafted blade-style 9-iron was assembled to an un-gripped length of 37.75 in (0.908 m), with loft and lie angles of 42° (0.733 rad) and 62.5° (1.091 rad), respectively. The shaft was a commercially available model, featured a tip diameter of 0.355 in (9.017 mm), a butt diameter of 0.600 in (15.24 mm), and was of 'stepped' design, i.e., the difference in diameter between the larger butt and smaller tip was achieved using steps, as opposed to a continuous taper. The club was fitted with a commercially available non-corded rubber golf grip. The test club was 'rigidly clamped' using a custom-built vice mechanism and positioned in front of a pneumatic ball cannon so that the clubhead was 600–630 mm from the nearest edge of the light gates, as illustrated in Figure 1. The point at which the club was clamped was representative of the gripping mechanism of the golf robot used at The R&A Equipment Test Centre; the robot gripping mechanism encloses approximately the first 0.23 m of the club as measured from the butt of the grip. The requirement of the gripping mechanism was solely that it needed to hold the club so that it could sustain multiple impacts.

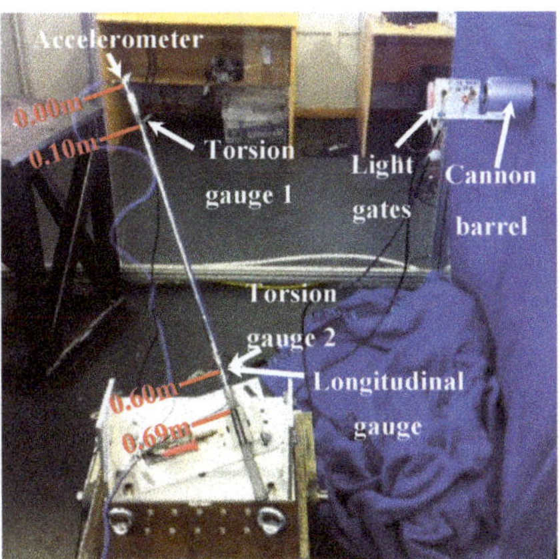

Figure 1. Experimental setup for strain measurement.

An accelerometer was positioned on the back of the clubhead to detect contact between ball and club. Three strain gauges were adhered to the shaft. The first was a gauge measuring torsion positioned on the shaft adjacent to the club's ferrule (approx. 0.1 m from the accelerometer measured parallel to the club shaft) on the 'underside' of the shaft, i.e., it would not have been visible to a golfer at address. A second gauge measuring torsion was located just beneath the grip (0.6 m from the accelerometer) on the same aspect of the shaft as the first gauge. A third gauge measuring longitudinal strain was located diametrically opposite to the second torsion gauge. Given that a fully-compensated circuit could not be achieved with the available resources, this third gauge offered reassurance that torsion was only being measured by the second torsion gauge, and that no cross-talk was occurring.

The accelerometer signal was split between two two-channel data acquisition devices to ensure synchronisation (the acceleration signal was used to trigger the measurement). The remaining channel of each device acquired a strain signal from one of the three available gauges. A third device was used independently to capture the signals from the light gates monitoring inbound ball speed. All data acquisition devices sampled at a frequency of 10 MHz. Strain signals were subsequently filtered using a third-order Butterworth filter; a cut-off frequency of 20 kHz was deemed appropriate after inspection of the signals in the frequency domain. It is acknowledged that this cut-off frequency may have been somewhat conservative; this was considered sensible however, given the aims of the study and lack of a fully-compensated strain gauge circuit.

Two target inbound speeds were used: 16 and 34 m·s^{-1}. The lower speed permitted a fidelity check in that the strains for this speed should be lower than those for the higher speed. The higher speed was selected to represent 9-iron clubhead speeds generated by elite male amateur golfers (approximately 38.9 m·s^{-1}) [19], however it was a slight underrepresentation as pilot tests with inbound speeds above 34 m·s^{-1} risked dislocation of the accelerometer from the back of the clubhead. A digital spirit level was used to quantify ($\pm 0.2°$) the orientation of the clubhead relative to the incoming trajectory of the ball. The shaft was angled such that the effective loft and lie of the clubhead were 26° and 0° respectively (i.e., the grooves were horizontal). Although this effective loft was slightly stronger (2–3°) than has been reported for 9-iron shots performed by elite amateurs [19], it was still considered to offer a realistic representation of clubhead presentation.

The clamping mechanism was then shifted so that the inbound ball would meet the club face at three different impact locations in turn: centre, toe and heel. The centre impact location was located at the geometric centre of the sandblasted area of the club face, whilst toe and heel locations were 17 mm towards the toe and heel of this centre point respectively (measured parallel to the grooves). The centre of gravity of the 9-iron clubhead featured in the study was actually 4 mm towards the heel and sole relative to the geometric centre, when measured in the plane of the club face. This meant that some rotation of the clubhead would be expected at all tested impact locations. Five repeat trials were performed at each of the six combinations of inbound speed and impact location.

2.3. Assumptions

The main assumption underlying the study was that the elastic reaction force at the gripping mechanism created a modified wave that returned along the shaft to the clubhead at the same speed as the initial wave. A linear superposition of the waves was assumed to be formed following this reflection, and as such interaction between wave types was not considered. Calculations and experimental results were based on a static club at impact; when used in a golf robot, the shaft would already be under a considerable amount of strain. However, these elastic stresses and strains ("pre-loading") would not be expected to affect material properties such as wave propagation speeds. The influence of the rubber grip of the club on strain propagation speeds was also considered negligible but it is noted that a rubber grip would introduce a strain-dependency to the club's response if strain waves travelled back to the clubhead before the end of the contact time between clubhead and ball.

3. Results

3.1. Inbound Ball Speed and Contact Time

The inbound ball speeds measured by the light gates were slightly lower than the target speeds, although, more importantly, they were consistent across impact locations. Mean ($\pm$SD) speeds when averaged across all impact locations were 15.6 ($\pm$0.4) and 33.5 ($\pm$0.6) m·s^{-1} for 16 and 34 m·s^{-1} target inbound speeds respectively. Variation in the centre of the impact location was inspected visually using a grid and found to be within 1 mm. Very little difference was found between the results recorded at the two speeds in terms of contact times. The broad initial peak in the accelerometer signals (Figure 2a) was approximately 0.5 ms in duration for both high and low tested inbound speeds. Contact time for a 9-iron impact does not therefore appear to be significantly different from results presented in previous studies concerning contact time measurement in golf [10–12].

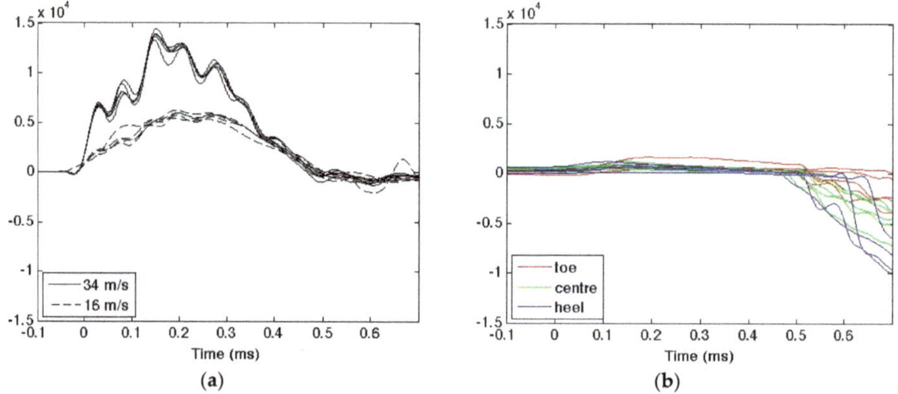

Figure 2. (a) Accelerometer signal recorded for five central impacts at both nominal inbound speeds; and (b) longitudinal strain signals (measured at the grip) for five trials at each of the three impact locations at nominal inbound speed: 34 m·s^{-1}.

In contrast, the amplitude of signals collected at 16 m·s^{-1} were smaller than those at 34 m·s^{-1} (Figure 2a). Results collected for the representative elite amateur level of 34 m·s^{-1} will be focused on in the future discussion.

3.2. Longitudinal Strain Measurement

Figure 2b shows longitudinal strain signals recorded for 34 m·s^{-1} impacts at centre, toe and heel impact locations. The signal appeared to demonstrate a very subtle departure from resting level at around 0.10 ms after initial contact, although more gross deviations were evident at after 0.50 ms. The latter could be attributable to bending modes: the second or third mode would be most probable in this case, considering their respective propagation speeds. The initial discontinuity occurred at a time reasonably close to that predicted (0.119 ms, see Table 1), although given the relatively small amplitude, this association is difficult to assert with a lot of certainty. These measurements did however offer reassurance that any meaningful strain, either longitudinal or bending, did not appear to be present at the grip until 0.5 ms when measured in this way.

Table 1. Theoretical (predicted) and measured timings (following initial contact) for longitudinal and torsional strain. Values in ms unless otherwise stated.

Location	Distance (m)	Longitudinal		Torsional	
		Predicted	Measured	Predicted	Measured
Tip	0.10	0.020	-	0.032	0.03–0.06
Grip	0.60	0.119	≈0.1	0.190	0.19–0.22
Clamp	0.69	0.137	-	0.218	-

3.3. Torsional Strain Measurement

Given that no significant variation in the longitudinal gauge occurred before 0.5 ms, it was considered safe to assume that anything detected earlier than this in the second torsion gauge (i.e., that situated at the grip) would in fact be torsion. The first indication of torsional strain is much clearer than for longitudinal measurements. It appears that torsional strain is first detected in the tip gauge at around 0.05 ms (Figure 3a), followed by the grip gauge at 0.20 ms (Figure 3b). This duration appears to become slightly shorter as the impact location moves towards the heel of the clubhead. The shorter durations measured for heel impacts agree very well with the predicted values (Table 1).

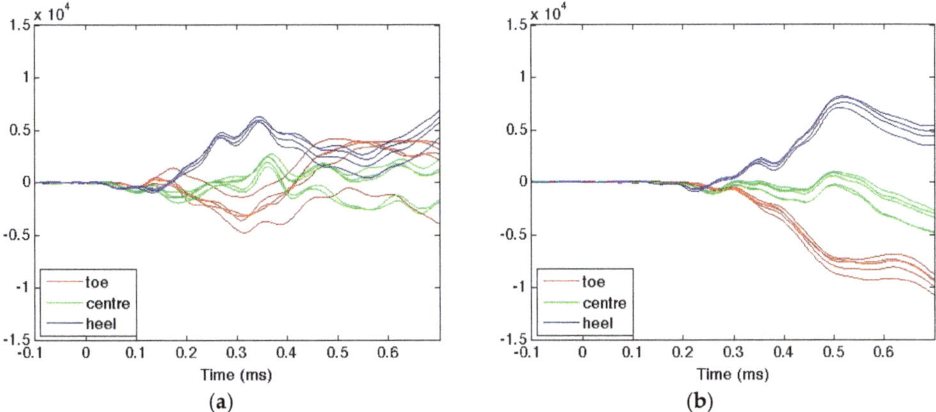

Figure 3. Torsional strain signals measured at the tip (**a**) and grip (**b**) for five trials at each of the three impact locations at 34 m·s^{-1}.

The positive vertical change in Figure 3a,b corresponds with closing of the club face, whilst a negative change is indicative of opening. The resultant torsion in the shaft therefore agrees with strain which would be expected, based on the current understanding of impact mechanics, following toe and heel impacts respectively, in that the club face opens following a toe impact, and vice versa.

4. Discussion

The critical duration in interpreting the results is double that of the predicted time for strain to reach the gripping mechanism, which assumes an instant reflection of the wave at this point. Doubling this duration accounts for the time taken for the wave to return to the clubhead, having been modified by the constraint at the clamp. If this total duration is longer than contact time between club face and ball, there is arguably no potential for the gripping mechanism to influence impact mechanics, and thus the outcome of the shot. Weaker gripping would make an effect even less likely as it would increase the time it takes until strain is reflected back from the gripping mechanism towards the clubhead.

Longitudinal strain was found to travel fast enough to have an effect (critical duration: 0.274 ms), but, as was stated in the methodology section, it is unlikely that deflections of this type will be large enough to be of any meaningful consequence, due to the shaft's stiffness in this direction. This viewpoint was vindicated by the very small deflections observed in the longitudinal strain gauge at around 0.1 ms after initial contact. The more gross deflections detected by the longitudinal gauge from 0.5 ms onwards were thought to be lower-order bending modes and were therefore too slow to have any influence on impact dynamics. Given their relative speeds, as were determined earlier using Equations (4) and (5), the first four modes would have taken 1.67, 0.66, 0.40 and 0.28 ms to reach the grip respectively. It is thought that this first mode is perhaps what was referred to in previous studies, which reported 1–2 ms as the time taken for such deflections to reach the grip [13,15,16].

Torsion was described as the strain that was of greatest concern, as per the discussion of theoretical wave speeds. Good agreement was found between measured and predicted timings, particularly for heel impacts and particularly when considering experimental error and scatter. The critical duration for torsional strain would be 0.436 ms, which indicates that the modified waves would return to the clubhead towards the very end of impact for a 9-iron shot. Their amplitude was thought to be great enough that they cannot be considered negligible (as was the case for longitudinal strain), however the degree of influence that they could exert in this last 0.050–0.075 ms of impact is questionable.

Furthermore, as irons become longer the critical duration will approach that of contact time. This means that as the potential for constrained clubhead motion to have a meaningful effect on ball launch conditions becomes even less as irons get longer, although the total length of club required for the critical duration to exceed 0.5 ms would be 1.023 m, based on the current experimental design (i.e., steel shaft, insertion depth of 0.23 m). To provide some context, this is approximately 1 cm longer than the longest iron (1-iron) based on reported industry averages [20], although both 1- and 2-irons are relatively uncommon in the modern game. Many commercial iron sets feature a 3-iron as the longest club, which is typically 3 in longer than a 9-iron taken from the same set, and thus the critical duration would be 0.483 ms.

5. Conclusions

Reasonable agreement has been shown between theoretically calculated strain propagation rates and experimental results (given experimental errors and scatter) collected using an instrumented steel-shafted 9-iron golf club. Longitudinal strain appears to propagate most quickly, but such deflections are likely to be small and therefore of little meaningful consequence. Shaft bending was not a primary concern of the study as modes of large enough amplitude appear to propagate too slowly to be relevant. Torsional strain propagates at a rate which suggests that constraints at the grip end of a golf club could potentially influence impact dynamics for steel shafted irons; however, this effect seems unlikely to be significant, a likelihood that decreases further for longer irons. As such, it is considered reasonable to accept the stated hypothesis and treat the influence of a robot's gripping mechanism on clubhead dynamics at impact as negligible, and therefore comparisons between robot and human data in this regard are valid. Similar consideration of shafts made from carbon composite materials would be worthwhile for future research, as would adoption of more sophisticated instrumentation that would allow direct comparisons between the influence of rigidly clamped constraints and human hands.

Acknowledgments: This research received partial funding support through a Postgraduate Student Award by the Department for Employment and Learning, Northern Ireland.

Author Contributions: T.W.C., E.S.W., M.S. and S.R.O. conceived and designed the experiments; T.W.C. performed the experiments, analysed the data and wrote the first draft of the paper; N.F.B., E.S.W., M.S. and S.R.O. reviewed the analysis and co-wrote the submitted paper.

Conflicts of Interest: The authors declare no conflict of interest.

References

1. Whittaker, A.; Thomson, R.; McKeown, D.; McCafferty, J. The application of computer-aided engineering techniques in advanced clubhead design. In Proceedings of the World Scientific Congress of Golf, St. Andrews, Scotland, 22–26 July 2002; Cochran, A.J., Ed.; E. & F. N. Spon: London, UK, 1990; pp. 286–291.

2. Milne, R.D.; Davis, J.P. The role of the shaft in the golf swing. *J. Biomech.* **1992**, *25*, 975–983. [CrossRef]

3. Harper, T. The Efficacy of a Golf Robot for Simulating Indvidual Golfer's Swings. Ph.D. Thesis, Loughborough University, Loughborough, UK, 2007.

4. Wicks, A.L.; Knight, C.E.; Braunwart, P.; Neighbors, J. Dynamics of a golf club. In Proceedings of the 17th International Modal Analysis Conference, Kissimmee, FL, USA, 8–11 February 1999; pp. 503–508.

5. Cochran, A.J.; Stobbs, J. *Search for the Perfect Swing*; Triumph Books: Chicago, IL, USA, 1968.

6. Daish, C.B. *The Physics of Ball Games*; English Universities Press: London, UK, 1979.

7. Iwatsubo, T.; Kawamura, S.; Miyamoto, K.; Yamaguchi, T. Numerical analysis of golf club head and ball at various impact points. *Sports Eng.* **2000**, *3*, 195–204. [CrossRef]

8. Harper, T.; Roberts, J.R.; Jones, R.; Carrott, A.J. Development and evaluation of new control algorithms for a mechanical golf swing device. *Proc. Inst. Mech. Eng. Part 1 J. Syst. Control Eng.* **2008**, *222*, 595–604. [CrossRef]

9. Roberts, J.R.; Harper, T.; Jones, R. Development of a golf robot for simulating individual golfer's swings. *Procedia Eng.* **2010**, *2*, 2643–2648. [CrossRef]

10. Johnson, S.H.; Lieberman, B.B. Experimental determination of apparent contact time in normal impact. In *Science and Golf IV*; Thain, E., Ed.; Routledge: London, UK, 2002; pp. 524–530.

11. Arakawa, K.; Mada, T.; Komatsu, H.; Shimizu, T.; Satou, M.; Takehara, K.; Etoh, G. Dynamic Contact Behavior of a Golf Ball During an Oblique Impact. *Exp. Mech.* **2006**, *46*, 691–697. [CrossRef]

12. Roberts, J.R.; Jones, R.; Rothberg, S. Measurement of contact time in short duration sports ball impacts: An experimental method and correlation with the perceptions of elite golfers. *Sports Eng.* **2001**, *4*, 191–203. [CrossRef]

13. Horwood, G.P. Golf shafts—A technical perspective. In *Science and Golf II*; Cochran, A.J., Farrally, M.R., Eds.; E. & F. N. Spon: London, UK, 1994; pp. 247–258.

14. Masuda, M.; Kojima, S. Kick back effect of club-head at impact. In *Science and Golf II*; Cochran, A.J., Farrally, M.R., Eds.; E. & F. N. Spon: London, UK, 1994; pp. 284–289.

15. Nesbit, S.M.; Hartzell, T.A.; Nalevanko, J.C.; Starr, R.M.; White, M.G.; Anderson, J.R.; Gerlacki, J.N. A Discussion of Iron Golf Club Head Inertia Tensors and Their Effects on the Golfer. *J. Appl. Biomech.* **1996**, *12*, 449–469. [CrossRef]

16. Whittaker, A. A study of the dynamics of the golf club. *Sports Eng.* **1998**, *1*, 115–124. [CrossRef]

17. Rossing, T.D.; Russell, D.A. Laboratory observation of elastic waves in solids. *Am. J. Phys.* **1990**, *58*, 1153–1162. [CrossRef]

18. Hall, D.E. *Musical Acoustics*, 2nd ed.; Brooks/Cole Pub. Co.: Pacific Grove, CA, USA, 1991.

19. Corke, T.W.; Betzler, N.F.; Wallace, E.S.; Otto, S.R. Clubhead presentation and spin control capability of elite golfers. *Procedia Eng.* **2013**, *60*, 136–142. [CrossRef]

20. Wishon, T.W. *The New Search for the Perfect Golf Club*; Fireship Press: Tuscon, AZ, USA, 2011.

Article

Measurement of Strain and Strain Rate during the Impact of Tennis Ball Cores

Ben Lane [1,2,3,*], Paul Sherratt [1], Xiao Hu [2] and Andy Harland [1]

[1] Sports Technology Institute, Loughborough University, Loughborough LE11 3TU, UK;
    p.j.sherratt@lboro.ac.uk (P.S.); a.r.harland@lboro.ac.uk (A.H.)
[2] School of Materials Science & Engineering, Nanyang Technological University, Singapore 637551, Singapore;
    asxhu@ntu.edu.sg
[3] Institute for Sports Research, Nanyang Technological University, Singapore 637551, Singapore
* Correspondence: b.lane@lboro.ac.uk; Tel.: +44-1509-564-827

Received: 28 January 2018; Accepted: 1 March 2018; Published: 4 March 2018

Abstract: The aim of this investigation was to establish the strains and strain rates experienced by tennis ball cores during impact to inform material characterisation testing and finite element modelling. Three-dimensional surface strains and strain rates were measured using two high-speed video cameras and corresponding digital image correlation software (GOM Correlate Professional). The results suggest that material characterisation testing to a maximum strain of 0.4 and a maximum rate of 500 s^{-1} in tension and to a maximum strain of -0.4 and a maximum rate of -800 s^{-1} in compression would encapsulate the demands placed on the material during impact and, in turn, define the range of properties required to encapsulate the behavior of the material during impact, enabling testing to be application-specific and strain-rate-dependent properties to be established and incorporated in finite element models.

Keywords: strain; strain rate; rubber; tennis; impact; digital image correlation

1. Introduction

Previous research assessing ball tracking data from professional tennis matches has established typical inbound impact conditions that the ball experiences, particularly when impacting court surfaces [1,2]. Knowledge of the impact conditions enables replication of typical impact conditions in the laboratory. Previous efforts to understand the behaviour of the ball core in isolation has focused, almost exclusively, on normal impacts with a rigid surface as a means of validating finite element (FE) simulations [3–7]. These simulations require characterisation of the mechanical behaviour of the rubber compound to predict the behaviour of the modelled ball core. Characterisation of ball core materials has so far been limited to quasi-static tensile, compressive, and stress relaxation testing. The issue with current test methods are the speeds and consequent achievable strain rates for testing. Typically, tensile tests have been conducted with a crosshead speed in the range of 50–999 mm·min^{-1} [5]. Elastomeric materials are known to exhibit strain-rate-dependent behaviour, whereby stiffness increases as strain rate increases, particularly at high strain rates [8–10]. When defining material behaviour for FE simulations, it is desirable to match the properties of the material when it is subjected to the strains and strain rates observed during use, especially if significant increases in stiffness are observed as strain rate increases.

Previous modelling techniques have utilised an iterative process, comparing artificial stiffening of stress–strain data and alterations to damping factors with high-speed video footage and force trace data of impacts to establish a material model providing a realistic representation of the ball during impact [3,5]. In part, the use of artificial stiffening is a result of the limitations of available material testing equipment. Only through simulation have maximum strains and strain rates during impact been estimated. Cordingley [3] estimated maximum strains in the region of 0.5 with maximum strain rates in the region of 300 s^{-1}. Allen [6] recorded maximum strains in the region of 0.3 for an impact against a rigid surface at 30 m·s^{-1}.

Traditional strain measurement devices such as strain gauges and extensometers do not lend themselves to high velocity ball impacts due to the associated connected electronics and possible alterations in deformation caused by the sensor being in contact with the surface. Optical strain measurement techniques utilising digital image correlation (DIC), such as GOM Correlate Professional (GOM GmbH, Braunschweig, Germany), provides a means of three-dimensional surface strain measurement utilising multiple cameras. GOM Correlate identifies stochastic image information, referred to as facets, by evaluating the grey values present within the facet. Optimal facets have a distinct and equally distributed pattern, with maximum possible contrast in grey values within the facet [11]. DIC could be employed to measure the strains and strain rates present during ball impacts and used to inform material characterisation testing, in turn enabling FE modelling to utilise, rather than techniques involving the artificial stiffening of quasi-static behaviour, material modelling techniques based on experimental data. Although not used to measure strain, GOM has been successfully implemented in a sporting context to measure golf ball and clubhead velocity [12].

The aim of this investigation was to establish the strains and strain rates experienced by tennis ball cores during impact, whereby the impact conditions are representative of professional play, in turn establishing the effectiveness of non-contact DIC for strain measurement during high-velocity, high-deformation sports ball impacts. A further aim was to define the range of strains and strain rates that material characterisation tests should incorporate to understand the material behaviour during use.

2. Materials and Methods

2.1. Equipment Set-Up

Ball cores were fired at a steel impact plate (15 mm thick) using a pneumatic cannon. The impact plate was positioned accordingly to achieve the desired impact angles, measured using a digital protractor. Two Photron Fastcam SA1.1 (Photron, San Diego, CA, USA) high-speed cameras were positioned approximately 0.75 m from the impact plate. The cameras were fixed to a beam, fitted to a tripod, and directed towards the centre of the impact plate, such that the angle between the cameras was approximately 24°. Light gates were positioned at the end of the barrel and connected to a counter, used to monitor ball speed and trigger video recording (Figure 1).

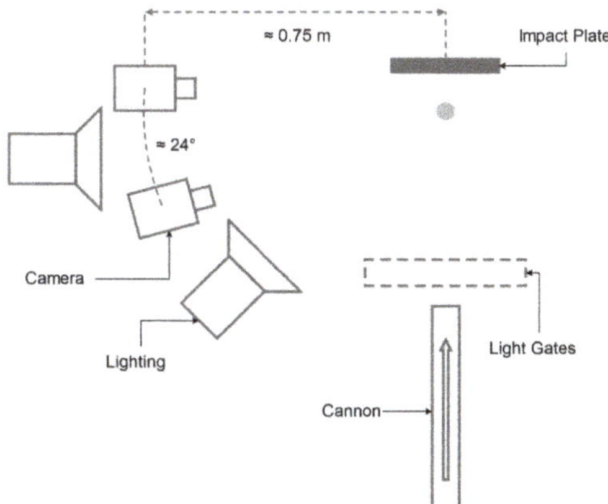

Figure 1. Plan view of lab set-up.

2.2. GOM

Both high-speed video cameras were fitted with 85 mm fixed focal length lenses (Zeiss planar T 1.4/85 mm ZF 2). The cameras operated as a pair to track the ball core during impact. The cameras were set-up for a capture volume around the impact plate of $275 \times 275 \times 275$ mm. The high-speed video cameras were synchronised and set to trigger as the ball passed through the light gates upon leaving the barrel of the cannon. Images were recorded at a frequency of 12,000 Hz, and the resolution was adjusted depending on impact angle to maximise pixel coverage of the ball (Table 1). Shutter speed ranged from 1/71,000 to 1/74,000 to prevent image blur while allowing sufficient light exposure. The aperture of the lenses was set at f/5.6, as recommended by GOM, and additional lighting, in the form of two flicker free ARRISUN 1200 W lights were utilised to resolve the problem of reduced light exposure.

Table 1. Camera settings for each impact angle.

Impact Angle	Shutter Speed	Image Resolution
18.9°	1/74,000	1024 × 496
27.6°	1/74,000	1024 × 496
40°	1/71,000	960 × 528
90°	1/74,000	1024 × 496

The volume of interest was calibrated in accordance with GOM instructions by capturing a series of images at full resolution (1024×1024 pixels) of a calibration object in multiple positions and orientations within the capture volume. The calibration object used was a GOM calibration panel, 200 mm × 160 mm in size. Calibration images were imported into GOM software (GOM Correlate Professional 2017, Braunschweig, Germany) and processed automatically given the specific details of the image sensor and calibration object used. A calibration file was produced and results given as to the quality of the calibration. As the cameras were repositioned to record at each impact angle, a separate calibration was performed for each, the results of which are given in Table 2. All calibrations were deemed successful as the quoted calibration deviation was less than 0.05 pixels, the maximum specified by GOM.

Table 2. GOM calibration results.

Impact Angle	Calibration Deviation	Measuring Volume
18.9°	0.020 pixels	275 × 275 × 275 mm
27.6°	0.020 pixels	275 × 275 × 275 mm
40°	0.019 pixels	275 × 280 × 280 mm
90°	0.019 pixels	280 × 280 × 280 mm

2.3. Details of Ball Core Preparation

Pressurised Wilson U.S. Open and pressureless Wilson Championship ball cores were used in the investigation. Twenty of each ball type were used, five balls per impact angle. Each ball was subjected to one impact per test speed, resulting in five impacts per core all at one given impact angle.

A stochastic pattern was applied to the cores using a black permanent marker to enable the GOM software to detect a surface and measure variables of interest (Figure 2). Due to the colour of the rubber compound of the pressureless ball cores, it was necessary to spray paint the cores matte white prior to applying the pattern. The contrast between the white paint and black permanent marker, comprising the pattern, was large enough for the GOM software to detect. The load-extension profiles for painted and non-painted samples highlighted the application of paint did not affect the stiffness of the rubber (Figure 3). The tests were conducted on a uniaxial test frame (Instron ElectroPuls E3000, Instron, High Wycombe, UK), which limited the test to quasi-static test speeds as high strain rate testing equipment was not available. Five repeats of each sample were tested (ISO 37 Type 1 sample size) at a speed of 400 mm·min^{-1}.

Figure 2. Ball preparation. Left: pressurised ball core; Right: pressureless ball core.

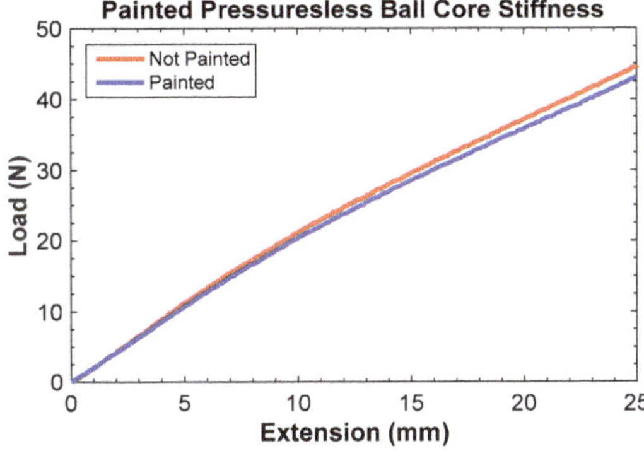

Figure 3. Effect of paint on the load-extension response pressureless ball core rubber (mean of five repeats displayed for painted and non-painted tensile samples).

Prior to testing, each ball core was subject to pre-conditioning as defined by the ITF ball approval process [13]. Each core was acclimatised at 20 °C and 60% humidity for 24 h prior to testing using a climate chamber (Alpha 190H, Design Environmental Ltd., Ebbw Vale, U.K.). The ball cores were then subjected to 9 pre-compressions by 2.54 cm sequentially through 3 perpendicular axes using an Instron ElectroPuls E3000 dynamic test system to remove set in the rubber.

2.4. Data Collection

In total 200 impacts were recorded, 100 each for each type of ball core. Four impact angles and five ball speeds were selected (Table 3), predominantly based on the findings of [1,2], to best replicate the impact conditions experienced during professional play. A maximum of five impacts per ball ensured minimal degradation in ball performance and, predominantly, ensured optimal pattern quality for the GOM software to process, as this could be inhibited by marks and smudges to the stochastic pattern. The nature of the pneumatic cannon meant the randomisation of ball speed to minimise order effects was not feasible. As a result, each ball was tested from the fastest test speed to the slowest. This was deemed the most suitable compromise as it ensured the stochastic pattern was in the best condition under the most testing conditions (least impact frames and largest deformation) for the GOM software to process. Image capture was triggered as the ball passed through the light gates. The high-speed videos were cropped to exclude frames where the ball was not present and saved in individual folders for each camera as 8-bit TIFF images.

Table 3. Impact condition selection.

Impact Angle (°)	Reasoning
18.9	Mean surface impact angle
27.6	Mean surface impact angle plus one standard deviation
40	Intermediate case between surface impact angles results and normal impacts
90	Simplified impact case, comparable to previous FE simulations
Ball Speed (m·s^{-1})	-
13.7	Mean pre-surface impact ball speed minus one standard deviation
22.1	Mean pre-surface impact ball speed
30.5	Mean pre-surface impact ball speed plus one standard deviation
43.1	Mean second serve ball speed (Men's)
52.5	Mean 1st serve ball speed (Men's)

2.5. Data Processing

Images were imported into GOM Correlate Professional 2017 where a surface component was created. Maximal surface coverage was achieved with a facet size in the region of 15 pixels and point difference in the region of 10 pixels, whereby a facet is a square section of the image and the point difference is the distance, in pixels, between the centre points of adjacent facets. The surface parameters were coherent with GOM recommendations whereby the point distance should be approximately two thirds the facet size. An image approximately five frames prior to impact was selected as the reference frame, against which all other strain calculations were compared. It was therefore assumed the ball core was under no strain at this point in time. Assuming zero strain was deemed valid due to the non-contact nature of the pneumatic cannon, subjecting the ball to minimal deformation on firing. Furthermore, on visual inspection and through use of GOM, no significant change in ball shape or surface strain was apparent prior to impact. Major and minor nominal strain, major and minor nominal strain rate and velocity were calculated in GOM and exported as CSV files. The frame of first and last contact with the impact plate were noted and used to define the frames of interest when analysing strain and strain rate.

Each CSV file contained the three-dimensional coordinates of each surface point alongside the strain, strain rate and velocity values for an individual frame. The number of surface points exported per file varied depending on how much of the surface defined in the reference frame was visible in that particular frame. Results were further processed in Matlab R2015b (Mathworks, Natick, MA, USA) to assess the frequency distribution of strain and strain rate during impact.

Major and minor strains were calculated by GOM using the principal axis transformation of the strain tensor. The principal axis transformation provided eigenvalues and eigenvectors for the new orthonormal basis of the strain tensor. The larger eigenvalue defined the major strain and the smaller eigenvalue defines the minor strain, with the direction of the strains defined by the corresponding eigenvectors (Figure 4). Major and minor strain rate were defined as the change in corresponding strain divided by the change in time.

To determine the maximum absolute strain for each element of the surface, the major and minor strains of the element were compared to determine the greatest absolute strain value. The value of greatest magnitude was selected for strain and the corresponding strain rate value was also selected based on the magnitude of the major and minor strain. The mean and standard deviation of strain and strain rate were calculated across all elements for all frames during impact, across all repeats. The frequency distribution of strain and strain rate was also assessed. The mean velocity of all elements of the surface prior to impact was used to define the pre-impact ball velocity.

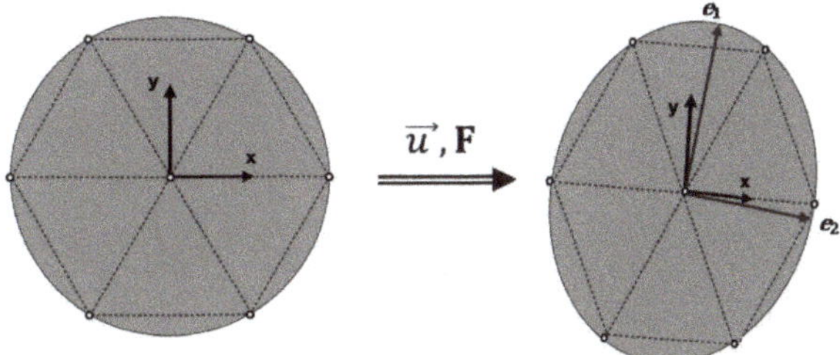

Figure 4. Major and minor strain definition, calculated using the eigenvectors (e1, e2) on deformation of the facet.

3. Results

To determine the ideal test conditions for material characterisation tests it was necessary to understand the strains and strain rates present under the most extreme impact conditions, to endeavour to capture the material behaviour within that range. Consequently, a select number of the impact conditions tested have been analysed in detail in this paper. The following subset of test conditions described in Table 3 were selected for analysis: 18° at 22.1 m·s^{-1}; 40° at 52.5 m·s^{-1}; 90° at 30.5 m·s^{-1}, 90° at 52.5 m·s^{-1}.

The largest mean strain during impact occurred for impacts at 40° and 52.5 m·s^{-1}, warranting further analysis. Similarly, the largest mean strain rate during impact occurred for impacts at 90° and 52.5 m·s^{-1}. Impacts tested at 90° with either of the two fastest impact speeds (43.1 m·s^{-1}, 52.5 m·s^{-1}) were found to lose surface tracking during impact, resulting in one or more frames where GOM was unable to compute a surface. Unlike impacts at 30.5 m·s^{-1} where the surface was tracked throughout impact duration (Figure 5). Loss of surface tracking occurred for both pressurised and pressureless ball cores due to the extreme levels of deformation, as shown in Figure 6. The 90° and 30.5 m·s^{-1} impact condition was selected to include impacts at 90° whereby the surface of the ball was tracked throughout impact as opposed to partially tracked. The impact condition representing the mean surface impact angle and ball speed found during professional play [1] was included to provide a benchmark against a more typical impact scenario.

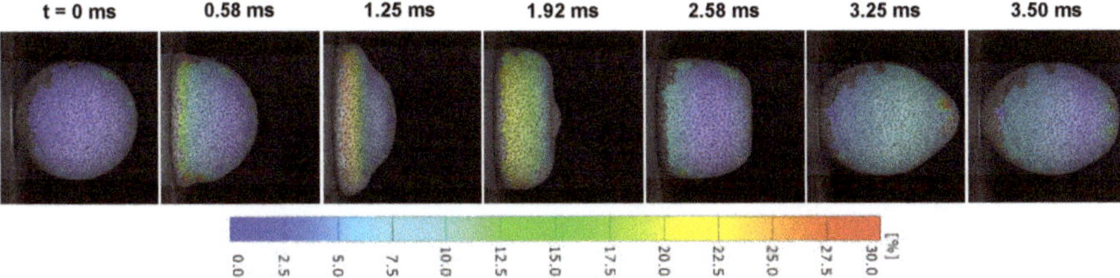

Figure 5. Surface coverage and major stain (%) during impact (30.5 m·s^{-1}, 90°). t = time in milliseconds.

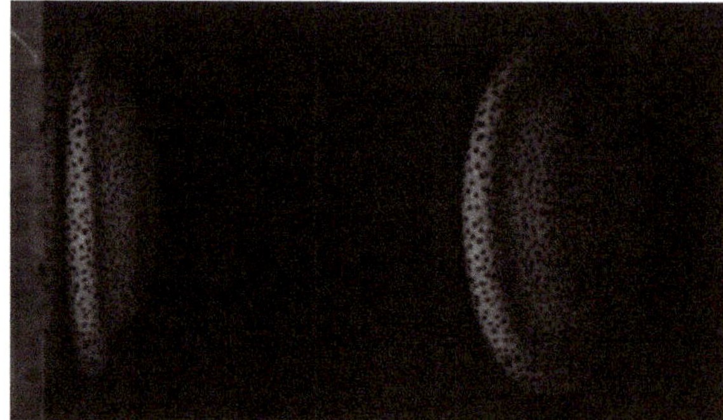

Figure 6. Example impact frames with loss of surface tracking for impacts at 52.5 m·s^{-1} and 90°.

3.1. Impact Ball Speed

Ball speed prior to impact was dependent on the air pressure setting within the pneumatic cannon system, rather than setting ball speed directly. The pressure within the system drifts resulting in alteration of the air pressure prior to each impact, consequently, the actual ball speed on impact varied. Mean ball speed for each speed and angle impact combination are given in Table 4. The maximum difference in mean ball speed to desired ball speed was −4.6 m·s^{-1} with a mean difference of −0.6 ± 1.4 m·s^{-1} when considering all impact conditions.

Table 4. Mean pre-impact ball speed ± one standard deviation (SD) for impact conditions of interest.

Angle (°)	Desired Ball Speed (m·s^{-1})	Pressurised	Pressureless
		Mean ± SD	Mean ± SD
18.9	22.1	22.9 ± 1.1	23.1 ± 1.2
40	52.5	54.8 ± 0.4	53.0 ± 0.6
90	30.5	30.4 ± 0.7	32.1 ± 1.2
90	52.5	52.1 ± 0.6	51.5 ± 1.7

3.2. Impact Strain

Figure 7 highlights that mean strain during impact increased with both impact angle and, to a greater degree, impact velocity. The exception being the mean strain for impacts at 90° with an impact velocity of 43.1 m·s^{-1} or greater which showed either no increase or a reduction in mean strain, a likely consequence of loss of surface computation at the largest deformations (e.g., Figure 6). The maximum mean strain occurred for impacts at 40° and 52.5 m·s^{-1} for both pressurised (0.06 strain) and pressureless ball cores (0.054 strain). Similarly, the minimum mean strain value occurred for impacts at 18.9° with in impact velocity of 13.7 m·s^{-1}.

Over 95% of strain measured was distributed between −0.1 and 0.1 strain for impacts at 18.9° with an impact velocity of 22.1 m·s^{-1} (Table 5). The remaining impact conditions at higher impact angles and faster impact speeds were more widely distributed (Figure 8). Table 5 also highlights a higher percentage of tensile (positive) strain is present than compressive (negative) strain, and was present to a greater degree at larger strains. For example, 8.4% of strain on average was distributed between 0.2 and 0.3 strain, compared to 2.1% for strains between −0.3 and −0.2, a trend present across all conditions except impacts at 18.9° and 22.1 m·s^{-1}.

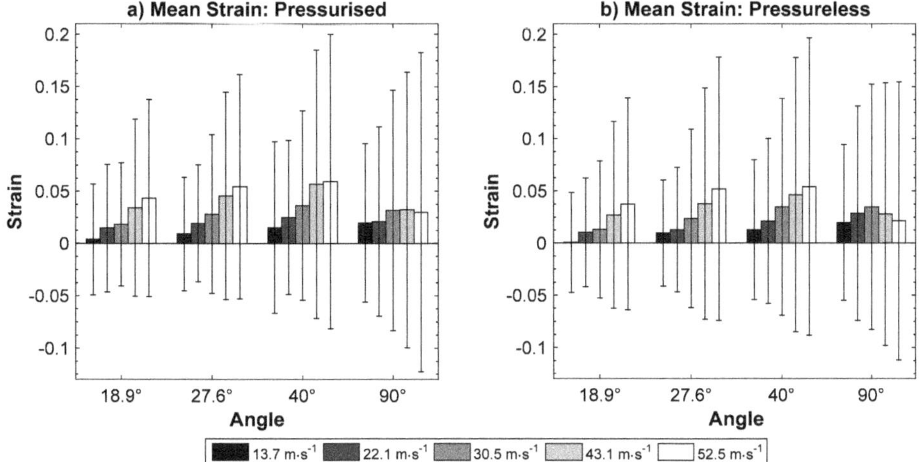

Figure 7. Mean strain during impact for pressurised (**a**) and pressureless (**b**) ball cores.

Table 5. Percentage distribution of strain during impact.

Strain Distribution		Pressurised Core: Impact Angle & Speed				Pressureless Core: Impact Angle & Speed			
Bin Edges-Strain		$18.9°, 22.1$ $m·s^{-1}$	$40°, 52.5$ $m·s^{-1}$	$90°, 30.5$ $m·s^{-1}$	$90°, 52.5$ $m·s^{-1}$	$18.9°, 22.1$ $m·s^{-1}$	$40°, 52.5$ $m·s^{-1}$	$90°, 30.5$ $m·s^{-1}$	$90°, 52.5$ $m·s^{-1}$
−0.4	−0.3	0.12	0.19	0.18	1.47	0.07	0.27	0.15	1.21
−0.3	−0.2	0.31	1.35	0.84	3.35	0.19	1.70	1.07	4.43
−0.2	−0.1	1.14	11.45	8.29	6.39	0.83	12.92	8.92	6.98
−0.1	0	34.59	25.11	35.97	32.13	49.08	24.69	32.38	30.24
0	0.1	61.11	21.19	30.72	34.77	47.35	18.35	32.04	35.34
0.1	0.2	2.02	25.35	13.98	11.10	2.06	27.80	14.55	12.92
0.2	0.3	0.34	9.58	8.94	5.79	0.20	10.23	9.97	6.09
0.3	0.4	0.10	5.27	0.84	4.04	0.09	3.66	0.72	2.47
0.4	0.5	0.04	0.26	0.08	0.42	0.03	0.16	0.08	0.12

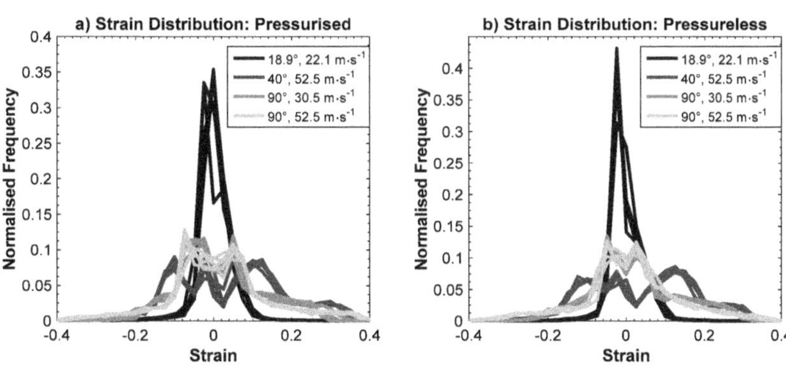

Figure 8. Strain distribution during impact for pressurised (**a**) and pressureless (**b**) ball cores.

3.3. Impact Strain Rate

In general, mean strain rate during impact increased in magnitude (in the negative direction) as impact speed at a given angle increased (Figure 9). Likewise, for a given impact speed the mean strain rate during impact increased in magnitude as the impact angle tended towards normal. Consequently, the mean strain rate of greatest magnitude occurred for impacts at $90°$ and 52.5 m·s^{-1} for both pressurised (-139 s^{-1}) and pressureless ball cores (-146 s^{-1}). The impact case representing the mean surface impact conditions ($18.9°$, 22.1 m·s^{-1}) was distributed evenly either side of zero. All other impact conditions were shifted in the negative direction, with negative strain rates accounting for around 60% of the distribution (Figure 10). No impact condition was found to have greater than 1% of the distribution whereby the strain rate was greater in absolute magnitude than 800 s^{-1} (Table 6).

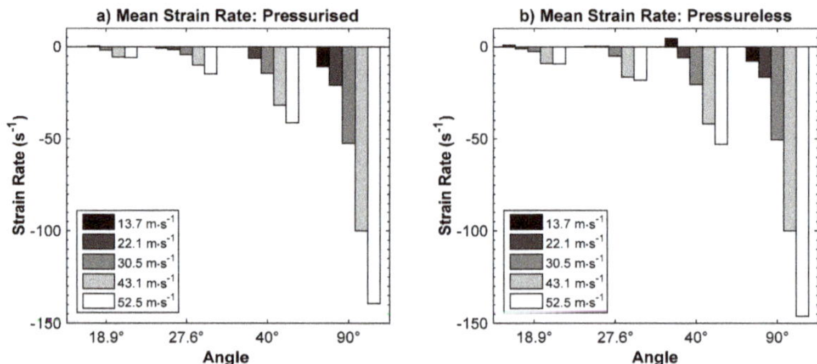

Figure 9. Mean strain rate during impact for pressurised (**a**) and pressureless (**b**) ball cores.

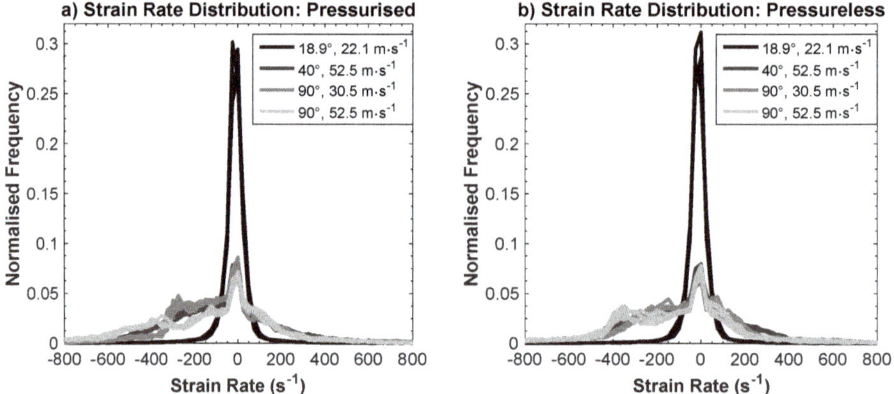

Figure 10. Strain rate distribution during impact for pressurised (**a**) and pressureless (**b**) ball cores.

Table 6. Percentage distribution of strain rate during impact.

Strain Rate Distribution		Pressurised				Pressureless			
Bin Edges		$18°, 22.1$ $\mathrm{m \cdot s^{-1}}$	$40°, 52.5$ $\mathrm{m \cdot s^{-1}}$	$90°, 30.5$ $\mathrm{m \cdot s^{-1}}$	$90°, 52.5$ $\mathrm{m \cdot s^{-1}}$	$18°, 22.1$ $\mathrm{m \cdot s^{-1}}$	$40°, 52.5$ $\mathrm{m \cdot s^{-1}}$	$90°, 30.5$ $\mathrm{m \cdot s^{-1}}$	$90°, 52.5$ $\mathrm{m \cdot s^{-1}}$
-800	-700	0.10	0.64	0.50	1.44	0.08	0.87	1.03	1.35
-700	-600	0.13	1.11	0.93	2.07	0.11	1.33	1.42	1.46
-600	-500	0.19	1.55	1.65	4.52	0.15	2.19	1.89	2.10
-500	-400	0.25	3.52	2.55	6.43	0.23	4.36	2.66	4.08
-400	-300	0.39	7.54	4.50	8.32	0.36	7.39	6.06	11.09
-300	-200	0.79	11.63	14.80	7.00	0.70	9.84	11.77	10.09
-200	-100	2.46	13.90	15.94	11.20	2.39	11.67	14.46	10.79
-100	0	44.01	18.34	20.46	15.31	45.33	17.38	17.31	15.96
0	100	47.19	17.33	17.11	17.83	46.76	18.02	19.09	18.15
100	200	2.00	8.62	8.87	9.27	2.00	9.36	10.38	7.91
200	300	0.75	5.01	4.39	3.84	0.67	6.32	4.29	3.08
300	400	0.38	2.94	2.10	1.82	0.33	3.43	2.14	1.71
400	500	0.24	1.75	1.66	1.39	0.20	1.31	1.70	1.26
500	600	0.19	1.01	1.30	0.97	0.12	0.87	1.43	1.01
600	700	0.12	0.74	0.83	0.52	0.07	0.62	1.13	0.64

4. Discussion

The results of this investigation highlight that both compressive and tensile strains were experienced during impact with a rigid plate. Regarding maximum strain values, tensile strains above 0.4 were uncommon, with frequency distributions highlighting that no frequency bin above a 0.4 strain contained more than 1% of the distribution. Consequently, material characterisation of ball core rubber in tension should test at a 0.4 strain to encapsulate the range seen during impact.

Experimental impacts of the pressurised ball core at 40° and 90° with an impact ball speed of 52.5 $\mathrm{m \cdot s^{-1}}$ had approximately 5% of strain distributed between 0.3 and 0.4, further evidence that 0.4 strain is the minimum desirable tensile strain for testing tennis ball rubber. These findings are slightly lower than the findings of Cordingley [3], who suggested that strains in the region of 0.5 can be observed during impact, which is more in keeping with the results of Allen [6].

Assessing the observed tensile strain rate measured by GOM highlighted that all impact conditions other than the mean surface impact condition (18°, 22.1 $\mathrm{m \cdot s^{-1}}$) had around 1% of the distribution between 500 and 600 $\mathrm{s^{-1}}$, no frequency bin above 300 $\mathrm{s^{-1}}$ accounted for more than 3.5%. Cordingley [3,4] reported maximum strain rates during ball core impacts of approximately 300 $\mathrm{s^{-1}}$. The results presented in this investigation indicated that tensile strain rates in the region of 600 $\mathrm{s^{-1}}$ were apparent, but to a limited degree (<1%). The ability to characterise the material at this rate would be optimal, but capturing material behaviour with a strain rate up to and including rates in the region of 500 $\mathrm{s^{-1}}$ would capture the vast majority of tensile strain rates experienced during impact.

Maximum observed compressive strains from GOM analysis suggested that strains up to -0.4 were present for impacts at 90° and 52.5 $\mathrm{m \cdot s^{-1}}$, accounting for around 1% of the distribution. Negligible amounts of strain (<0.2%) were measured for compressive strains greater in magnitude than -0.4. The results suggest that material characterisation in compression would encapsulate the vast majority (>95%) of observed compressive strain during impact if tested to strains of -0.4, whilst testing to strains of -0.3 may be sufficient depending on the severity of the impact conditions.

Only impacts at 90° and 52.5 $\mathrm{m \cdot s^{-1}}$ had more than 1% of the distribution greater in magnitude than -800 $\mathrm{s^{-1}}$. The remaining impact conditions highlighted a maximum compressive strain rate closer to -700 $\mathrm{s^{-1}}$. Frequency bins between -400 and -800 $\mathrm{s^{-1}}$ for the 90°, 52.5 $\mathrm{m \cdot s^{-1}}$ impact condition account for more than double the percentage of the other impact conditions. Missing frames or obstructed surface coverage likely resulted in an overrepresentation of results between -400 and -800 $\mathrm{s^{-1}}$. These results appear to be consistent with results in the literature. Allen [6] highlighted a maximum strain for simulation at 30 $\mathrm{m \cdot s^{-1}}$ of 0.3, consistent with GOM results for 90° and 30.5 $\mathrm{m \cdot s^{-1}}$, whereby all frequency bins above a 0.3 strain contained less than 1% of the distribution.

GOM distributions for the mean surface impact test case (18°, 22.1 m·s^{-1}) appear to follow a normal distribution, but the remaining cases do not. For strain, these distributions have multiple peaks rather than one distinct peak around zero, whilst the strain rate distributions are skewed in the negative direction, with negative strain rates accounting for around 60% of the distribution.

The strain rates presented in this investigation are in the 100–1000 s^{-1} order of magnitude, at which strain-rate-dependent behaviour has been observed in vulcanised rubbers [8–10]. To test at rates in this order of magnitude, a bespoke test rig (similar to that used by Hoo Fatt & Bekar [9]) or more complex testing equipment, such as a split Hopkinson pressure bar, is required. Characterisation at strain rates between 100–1000 s^{-1} would confirm if this rate-dependent behaviour is present in ball core rubber under the conditions experienced during use. Should significant rate dependence be observed, the incorporation of more complex material models (compared to hyperelastic models) incorporating this behaviour can be utilised in finite element models of ball impacts. Furthermore, should minimal strain rate dependence be observed, the process of artificial stiffening of hyperelastic material models to represent ball core rubber, as utilised previously [3–5], should no longer be employed. Additionally, different material compounds can be compared at representative strain and strain rates when assessing potential new compounds for use as ball core rubber.

GOM is limited by the surface coverage, which is dependent on camera positions and relative placement. Clearly with one camera pair, it is not possible to obtain full surface coverage, particularly as the underside of the ball is obscured by the impact plate. Furthermore, it is only possible to measure parts of the object that are visible, so strains through the thickness of the material and on the inside surface of the cavity cannot be measured.

Increased ball coverage could be achieved using additional cameras. A calibration of each camera 'pair' would be required in addition to the calibration of a global reference frame, to which all surface analysis could be transposed and combined to achieve more complete coverage. The time required to capture and process the data would greatly increase, however, and access to multiple high-speed cameras with each pair of cameras requiring matching lenses would be required.

5. Conclusions

DIC has been shown to be a useful tool in determining the strains and strains rates present during the impact of tennis ball cores against a rigid surface; however, at the upper limits (52.5 m·s^{-1}, 90°), deformations can become too large to maintain surface coverage.

The results suggest that material characterisation testing to a maximum strain of 0.4 and a maximum rate of 500 s^{-1} in tension and a maximum strain of -0.4 and a maximum rate of -800 s^{-1} in compression would encapsulate the demands placed on the material during impact.

Knowledge of the strains and strain rates present during impact allows for material characterisation under these conditions. In turn, material modelling techniques capable of capturing the strain-rate-dependent behaviour can be employed. Furthermore, the surface strain analysis can be utilised as a means of validating FE simulations of the ball core.

Acknowledgments: The authors would like to thank Wilson Sporting Goods for supplying the ball cores used in this study.

Author Contributions: Ben Lane, Paul Sherratt, Xiao Hu and Andy Harland were involved in conceiving and designing the experiments. Ben Lane performed the experiments. Ben Lane and Paul Sherratt analysed the data and wrote the paper.

Conflicts of Interest: The authors declare no conflict of interest.

References

1. Lane, B.; Sherratt, P.; Hu, X.; Harland, A. Characterisation of ball degradation events in professional tennis. *Sports Eng.* **2017**, *20*, 185–197. [CrossRef]
2. Lane, B.; Sherratt, P.; Xiao, H.; Harland, A. Characterisation of ball impact conditions in professional tennis: Matches played on hard court. *Proc. Inst. Mech. Eng. Part P J. Sports Eng. Technol.* **2015**, *230*, 236–245. [CrossRef]
3. Cordingley, L. *Advanced Modelling of Surface Impacts from Hollow Sports Balls*; Loughborough University: Loughborough, UK, 2002.
4. Cordingley, L.P.; Mitchell, S.R.; Jones, R. Measurement and modelling of hollow rubber spheres: Surface-normal impacts. *Plast. Rubber Compos.* **2004**, *33*, 99–106. [CrossRef]
5. Sissler, L. *Advanced Modelling and Design of a Tennis Ball*; Loughborough University: Loughborough, UK, 2012.
6. Allen, T.B. *Finite Element Model of a Tennis Ball Impact with a Racket*; Sheffield Hallam University: Sheffield, UK, 2009.
7. Goodwill, S.R.; Kirk, R.; Haake, S.J. Experimental and finite element analysis of a tennis ball impact on a rigid surface. *Sports Eng.* **2005**, *8*, 145–158. [CrossRef]
8. Bergström, J.S.; Boyce, M.C. Large strain time-dependent behavior of filled elastomers. *Mech. Mater.* **2000**, *32*, 627–644. [CrossRef]
9. Fatt, M.S.; Bekar, I. High-speed testing and material modeling of unfilled styrene butadiene vulcanizates at impact rates. *J. Mater. Sci.* **2004**, *39*, 6885–6899. [CrossRef]
10. Roland, C.M. Mechanical behavior of rubber at high strain rates. *Rubber Chem. Technol.* **2006**, *79*, 429–459. [CrossRef]
11. GOM. *Digital Image Correlation and Strain Computation Basics*; GOM: Braunschweig, Germany, 2016.
12. Leach, R.J.; Forrester, S.E.; Mears, A.C.; Roberts, J.R. How valid and accurate are measurements of golf impact parameters obtained using commercially available radar and stereoscopic optical launch monitors? *Measurement* **2017**, *112*, 125–136. [CrossRef]
13. International Tennis Federation (ITF). *ITF Approved Tennis Balls, Classified Surfaces & Recognised Courts*; ITF: London, UK, 2017.

Article

Table Tennis Ball Impacting Racket Polymeric Coatings: Experiments and Modeling of Key Performance Metrics

Renaud G. Rinaldi [1,*], Lionel Manin [2], Sébastien Moineau [3] and Nicolas Havard [3]

[1] MATEIS, CNRS UMR 5521, INSA-Lyon, Univ Lyon, F-69621 Villeurbanne, France
[2] LaMCoS, CNRS UMR 5529, INSA-Lyon, Univ Lyon, F-69621 Villeurbanne, France; lionel.manin@insa-lyon.fr
[3] Cornilleau, Bonneuil-Les-Eaux, 60121 Breteuil, France; smoineau@cornilleau.fr (S.M.); nhavard@cornilleau.fr (N.H.)
* Correspondence: renaud.rinaldi@insa-lyon.fr; Tel.: +33-04-72-43-62-09

Received: 7 November 2018; Accepted: 29 December 2018; Published: 4 January 2019

Abstract: The performance of a table tennis racket is often associated with subjective or quantitative criteria such as the adhesion, the control and the speed. Overall, the so-called performance aims at characterizing the impact with the ball. Ultimately, the polymeric layers glued onto the wooden blade play a key role, as evidenced in a previous work where the normal linear (no spin) impact of a ball onto polymeric layers was experimentally and numerically investigated. In this work, more realistic loading conditions leading to varying the incident angle and spin of the ball, were explored. While the sole linear restitution coefficient was determined in the anterior normal impact study, new physical metrics were identified to describe fully the trajectory of the reflected ball after impact. A companion 3D finite elements model was developed where the polymeric time-dependent dissipative compliant behavior measured with dynamic mechanical analysis and compression tests was accounted for. The confrontations with the experimental data highlighted the key role of the polymer intrinsic properties along with the friction coefficient between the ball and the polymer external layer.

Keywords: impact; polymer; rate dependence; architecture; friction; finite elements

---

1. Introduction

A table tennis racket comprises two polymeric sheets, the foam and the compact that are glued together and on to a wooden paddle. Both polymeric constituents are in their rubbery state at ambient temperature so that they exhibit large and mostly instantaneous reversible deformation when impacted by a ball. Aside from preventing the ball from breaking, these compliant layers trigger the speed and control of the ball and the identification of the design parameters controlling the overall performance is of great importance. Looking at the complex physical phenomenon taking place when the ball impacts a polymeric pad, the friction coefficient, the polymeric intrinsic bulk properties and the layers geometries and architectures correspond to the many potentially adjustable parameters. That being said, when considering the impact problem attention can be focused on the impactor, the target and/or the contact between the two.

While many studies were devoted to studying the kinematic behavior (trajectory and aerodynamics) of a thin shell hollow sphere impacting a flat surface, most of these works used rigid surface, thus being representative of the ball/table impact [1–7]. When the rebound of the ball was the focus of attention, the coefficient of restitution, corresponding to the ratio between the incident and reflected linear velocities could be calculated [2,8]. This parameter only relied on the kinematic description of the impact and on global dynamics so that little to no attention was often given to the materials description. In the end, the coefficient of restitution metric characterized the overall speed performance during impact, which corresponded to the performance of a given combination of impactor and target as a whole.

Impact under normal incidence (and no spin) was often employed to further study the deforming behavior of the ball when in contact with a rigid surface. For both the quasi-static and the dynamic loading regime, the reversible buckling of the thin-wall sphere has been observed and modeled [9–12]. Here, further attention was given to the material mechanical properties. Nevertheless, when used as the model material, the table tennis ball's celluloid constitutive behavior could be described as an elastic perfectly plastic material [11] even though polymers are known to be dissipative rate-dependent materials [13,14]. Recently, Zhang et. al. used an inverse-fitting method to describe the visco-elastic properties of the celluloid material and to model the impact of a table tennis ball impacting a rigid surface, evidencing the need for the account of the polymeric materials rate-dependency [8].

The impact of tennis table ball on a racket with coverings had been thoroughly studied by Tiefenbacher et al. [15] from a macroscopic analysis, both numerically and experimentally. The impact was analyzed based on two restitution coefficients (E_{par}, normal ball velocity, T_{par} tangential ball velocity). It was shown that the tangential elasticity of the coverings had a high influence on these coefficients, the authors identified a so called 'tangential effect'. The details of the ball—covering contact during impact were not discussed but only analyzed through the previous coefficient of restitution that were determined for numerous coverings and impact conditions (velocity, angle and spin frequency). Following that study, Tiefenbacher et al. [16] had analyzed the influence of special equipment materials on decisive strokes.

The contact conditions during the impact was studied by Cross [17], but for tennis, golf, basketball and baseball balls. The author based his analysis on the measurement of the tangential and normal forces during the ball–surface impact. The tangential force being seen as a parameter that permit identifying the contact condition: sliding, rolling or gripping of the ball on the impacted surface. The author highlighted that depending on its incidence angle and its linear speed and spin frequency, the ball could grip or slide but did not roll on the impacted surface. However, the author did not consider table tennis balls which behave differently from tennis balls, exhibiting for instance a lower mass/volume ratio.

Finite element modelling (FEM) of sports balls impacting rackets, bats, clubs or surfaces has been considered by many authors in the last two decades. Goodwill, Haake and Allen [18–21] extensively investigated the impact of a tennis ball on a rigid surface or on a string bed, studying the contact conditions of the tennis ball with the impacted surface and the reversal of spin. More generally, depending on the sport under investigation, noticeable differences exist between the two impacting bodies. They can be considered rigid, soft, plain or hollow so that the deformations experienced by the ball and/or the target can be large and/or small. Ultimately the corresponding models have to be adapted to these specificities. The constitutive equations, often being nonlinear, has to be implemented in the FEM, requiring most of the time the experimental characterization of the material parameters. Another key point for getting a consistent modeling is the contact formulation and its characteristics: friction coefficient, contact state (stick/slip).

In a recent work, Rinaldi et al. studied the impact response of a ball onto targets corresponding to the polymeric layers of a table tennis paddle [22]. Normal impacts with varying incident linear velocities and no spin were solely considered and joint experimental measurements and 3D FEM were performed. The linear coefficient of restitution corresponded to the quantitative metric chosen for comparison and fair agreement was obtained. The buckling of the ball was confirmed numerically meaning that the sample/ball contact zone corresponded to a ring with its mean radius increasing till the maximum crushing. It is worth noting that the material properties, the geometry and the local contact zone were considered in the model, so that parametric studies allowed to evidence their importance. Nevertheless, the explored impact conditions were away from realistic conditions where angular velocity (ball spin) and non-normal incidence usually prevail.

Thus, the present research aimed at exploring an enhanced set of incident loading conditions, as well as challenging the finite element model in matching the experimental observations. Along with the large set of impact tests, new metrics had been defined, monitored and confronted.

2. Materials and Methods

2.1. Materials

The balls and the polymeric pads were provided by the French company Cornilleau©. Section views of the three elements in the pristine state, and reconstructed from micro Computed Tomography (μ-CT) measurements, are presented in Figure 1, illustrating the geometry, architecture and microstructure of the different constituents.

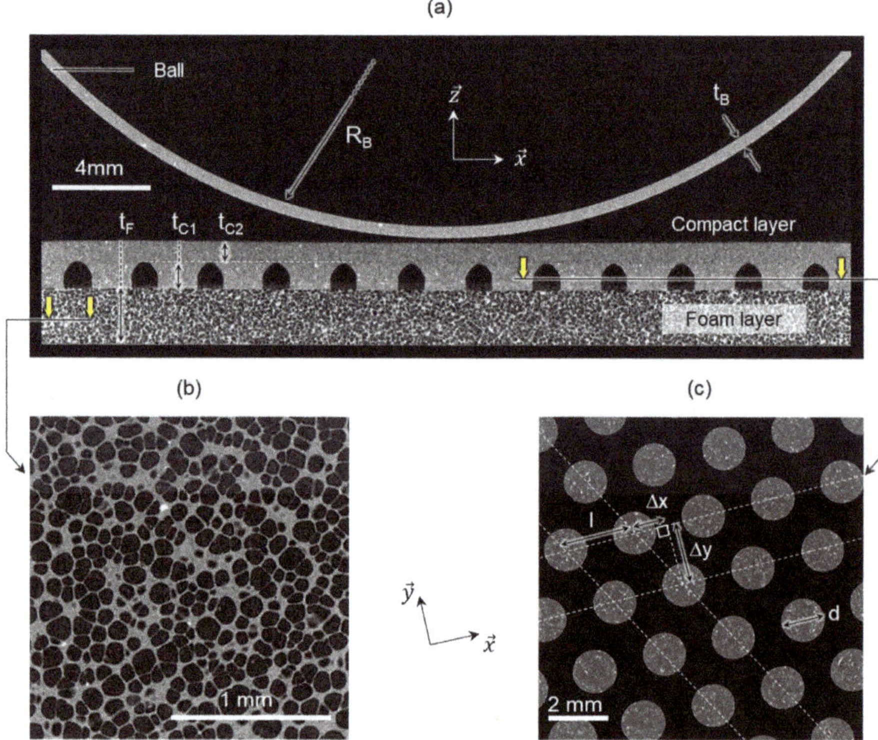

Figure 1. X-ray tomography measurements (**a**) 2D X-ray tomography imaging of the ball and foam + compact pad (Voxel size = 17 μm). (**b**) Section view (xz) of the foam microstructure (Voxel size = 2.5 μm). (**c**) Section view (xy) highlighting the periodic pattern of the circular pimples (Voxel size = 17 μm).

Key geometrical parameters and their values are listed in Table 1. They were obtained from the image analysis of the 3D μ-CT data or mentioned otherwise. It is worth noting that the pimples diameter varied throughout the thickness (z axis) and that only the mean value was reported.

Table 1. Dimensions of the ball and polymer layers.

Description	Symbol	Dimension
Ball		
Radius	R_B	20 mm
Wall thickness	t_B	0.39 mm
Foam		
Thickness	t_F	2 mm
Relative density [1]	$\underline{\rho}_F$	0.44
Density [2]	ρ_F	490 kg/m^3

Table 1. *Cont.*

Description	Symbol	Dimension
Compact		
Thickness of the pimples	t_{C1}	0.96 mm
Thickness of the dense layer	t_{C2}	0.81 mm
Spacing	l	2.37 mm
Offset along the x axis	$\triangle x$	2.08 mm
Offset along the y axis	$\triangle y$	1.18 mm
Density of the parent material [2]	ρ_F	1000 kg/m^3

[1] determined via thresholding and post-processing the stack of μ-CT images with ImageJ©. [2] determined via measuring and weighting large samples (100 m × 10 mm × total thickness).

2.2. Methods and Samples Preparation

A testing apparatus property of Cornilleau© and illustrated in Figure 2a, had been used to launch table tennis balls horizontally, while controlling the ball's incident linear and spin velocities. The targets consisted of 100 mm^2 squares cut from the paddle coverings where the compact and the foam layers were separated if needed. The two diagonals were drawn on the square surface, as shown in Figure 2b, in order to visually ascertain the ball impact within the center region of the targets.

The squared samples were taped on to a wooden plate that was screwed on the rigid frame, 260 mm away from the launcher's end. The target's frame could be rotated around the x-axis so that the incident angle between the ball and the paddle could be controlled. The ball's outside surface was marked with meridians (see Figure 2c): a circular dashed line allowed location of the seal and two solid lines were drawn using a mask. The dashed line was used to visually detect and exclude from the post-processing routine the tests where impact occurred on the seal. Also, the solid lines were used by the in-house post-processing software to dissociate the ball's linear and angular velocities (see Section 2.3).

A camera with long exposure time settings and stroboscopic lights was used to take one image per impact test. The obtained image displayed the target as well as numerous ball prints before and after impact as evidenced in Figure 3a. Consequently, the time Δt, related to the strobe frequency f (Δt = 1/f) elapsed between two consecutive traces. It is worth mentioning that the principle is similar to that used by Carre et al. [23].

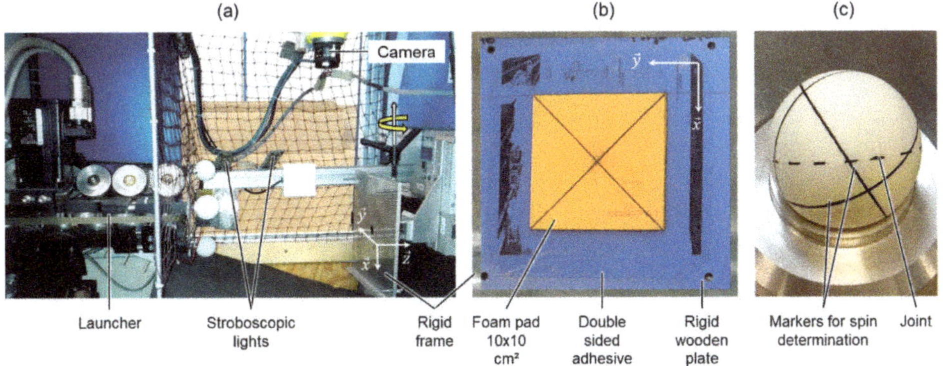

Figure 2. Impact tests. (**a**) Cornilleau home-made apparatus. (**b**) Top view of a foam pad glued onto a wooden plate frame. (**c**) Ball with visual markers drawn.

Ten impacts were performed to characterize a given set of *{Ball - Target - testing conditions}* so that error bars corresponding to the standard deviation were added to the plots displaying the experimental results. The measured variabilities apply to both the incident and reflected trajectories and correspond to the error bars along the x- and y-axis respectively in the coming figures. Finally, a target corresponded to one pad and the pad thickness was measured prior and after testing to ascertain no substantial damage/wear.

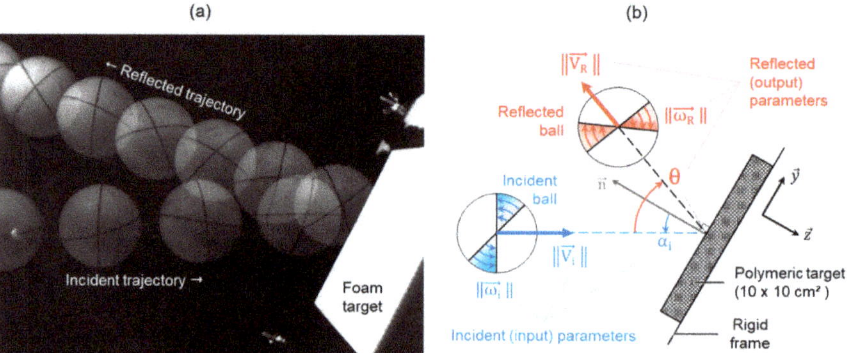

Figure 3. (**a**) Raw image of an impact test showing traces of the ball before and after contact. (**b**) Schematization of the kinematic parameters describing the motion of the table tennis ball prior (incident) and after (reflected) impact.

2.3. Image Post-Processing and Metrics of Performance

Post-processing software, property of Cornilleau©, was employed to determine the ball's complex trajectory from the obtained images. The non-orthogonal markers (Figure 2c) were recognized on the traces so that the 2D trajectory could be reconstructed [23,24]. Six kinematic parameters were finally monitored. They are displayed in Figure 3b. Three parameters were related to the incident trajectory:

- α_i is the incident angle, in degrees, calculated from the normal to the target surface $\vec{n}$.

- $\|\overrightarrow{V_i}\|$ (also denoted V_i in the following) corresponds to the magnitude, in meter per second (m/s), of the incident linear velocity. Considering the experimental setup and coordinate system defined in Figure 3b, $\overrightarrow{V_i}$ had components along the in-plane y- and z-axes,

- $\|\overrightarrow{\omega_i}\|$ (also denoted ω_i in the following) corresponds to the magnitude, in revolution per second (rev/s), of the incident angular velocity (spin). Here, $\overrightarrow{\omega_i}$ was collinear with the out of plane *x-axes*.

Three parameters fully defined the reflected trajectory:

- θ is the deviation angle in degrees, calculated from the incident orientation. Clockwise angles were set positive.

- $\|\overrightarrow{V_R}\|$ corresponds to the magnitude, in meter per second (m/s), of the reflected linear velocity. $\overrightarrow{V_R}$ had components along the in-plane y- and z-axes,

- $\|\overrightarrow{\omega_R}\|$ corresponds to the magnitude, in revolution per second (rev/s), of the reflected angular velocity (spin). $\overrightarrow{\omega_R}$ was collinear with the out of plane x-axes,

Two assumptions accompanied the previous definitions. First, the ball was supposed to remain in the horizontal (xy) plane, suggesting that the effect of gravity was negligible. It was confirmed by verifying (using the circle fit functionality in the image processing software imageJ©), that the areas of the ball traces remained constant in one image. Indeed, if the camera were fixed on top of the set-up, the gravity would have contributed to out of plan motion and resulted in smaller traces of the ball on the image. Second, both the incident and reflected trajectories were assumed straight (at least between two consecutive traces), neglecting potential aerodynamics effects.

With the trajectories fully defined, quantitative metrics were set in order to quantify the performance of the impacted polymeric layers. In the previous study focusing on normal impact with no spin [22], the sole linear coefficient of restitution was considered, as defined below:

$$CR_L = \frac{\|\vec{V_R}\|}{\|\vec{V_i}\|},\tag{1}$$

CR_L is unitless, positive, smaller than 1 (when the target is fixed) and related to the speed performance. CR_L remains a relevant metric and was used in this study. The broader spectrum of impact conditions explored suggested the use of two other metrics, namely the deviation angle θ and the (unitless) spin ratio SR defined as:

$$SR = \frac{\vec{\omega_R}}{\vec{\omega_i}},\tag{2}$$

As mentioned above, both incident and reflected spin axes were collinear with the out of plane *x-axis*. Thus, the spin ratio SR is a real number with absolute value that can be greater than 1. The spin ratio was also studied by Allen et al. [25] in the case of tennis balls. Thus, the aforementioned three parameters were related to the ball's trajectory and aimed at characterizing the performance of the impact between a given *{Ball - Target}* combination.

3. Experimental Results

Figure 4 presents, in a double entry array form, an overview of the impact conditions tested on the foam sample. Impact tests with varying negative (anticlockwise) angular velocities and incident angles are reported in the different rows and columns respectively. All the tests were performed at a constant linear velocity of 14 m/s and ambient temperature. Three different spin (0, −33 and °66 rev/s) and three different angles (0, 30° and 60°) were considered. The top left image corresponds to a normal impact with no spin that has been studied and documented in the previous study [22]. Under the new tested loading conditions, incident and reflected trajectories were easily distinguished.

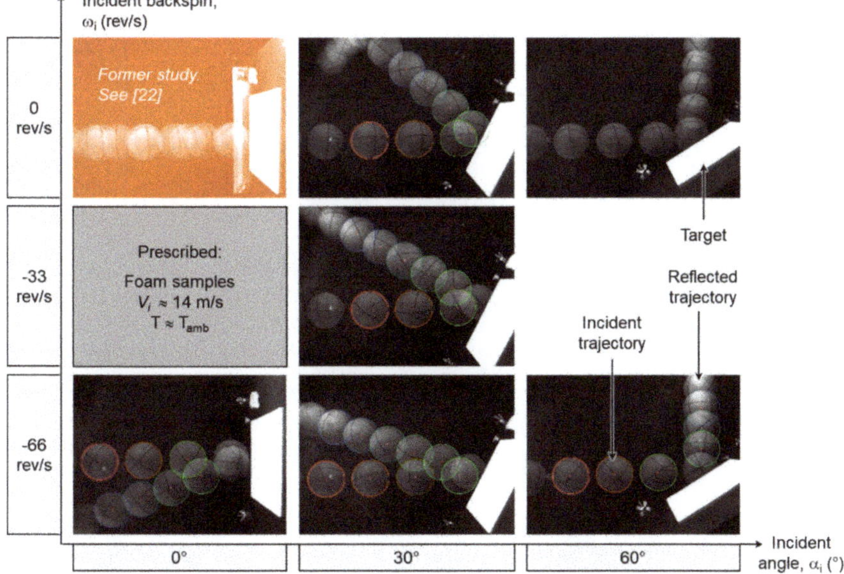

Figure 4. Raw images of the various impact tests performed on the foam sample.

Studying the effect of the incident spin for a fixed angle, or the effect of the incident angle for a fixed spin was obtained by focusing on one row or column respectively. For instance, the effect of the incident angle when the ball was thrown at 14 m/s with a backspin of 66 rev/s is presented in Figure 5. Figure 5a displays the raw images of the impact performed on the *foam + compact* (row 1) and *foam* (row 2) samples. At first sight, noticeable differences in the reflected trajectories was observed between the two samples for a given $\{\omega_i;\alpha_i\}$.

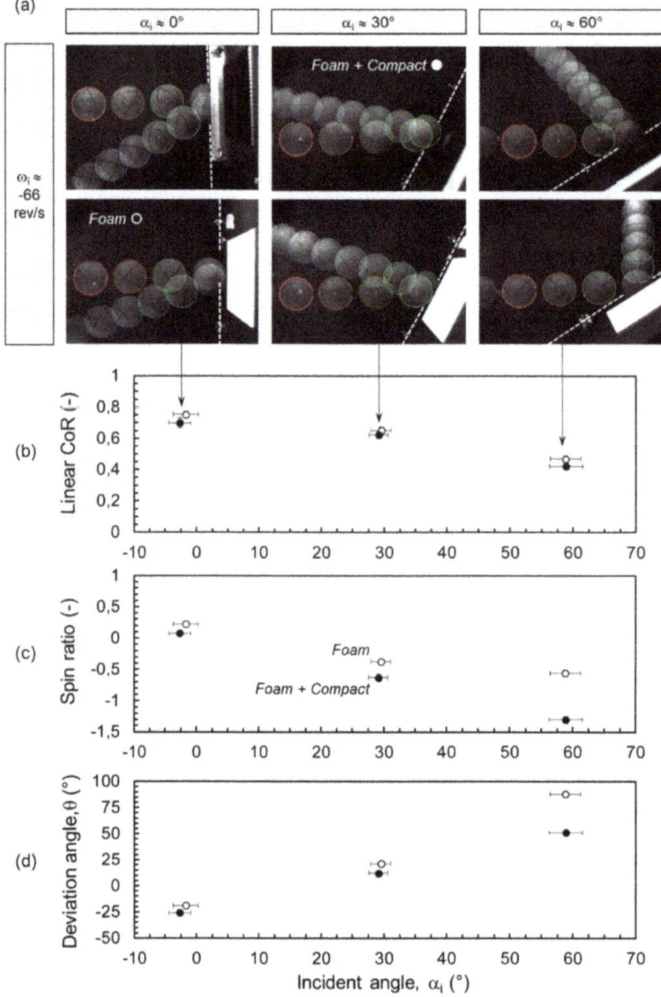

Figure 5. (**a**) Raw images of impact tests performed on the *foam + compact* and *foam* samples with varying incident angle ($\|\vec{V_i}\| = 14$ m/s, T_{amb}). (**b**) Linear coefficient of Restitution as a function of the incident angle. (**c**) Spin ratio as a function of the incident angle. (**d**) Deviation angle as a function of the incident angle.

Further quantitative analysis is presented in Figure 5b–d where the linear coefficient of restitution, the spin ratio and the deviation angle are respectively plotted as a function of the incident angle. Increasing the incident angle led to a decrease in the linear coefficient of restitution, the values for the *foam + compact* being slightly smaller than those measured on the *foam* alone.

For the two other metrics, a change in the sign was observed between the normal tests ($\alpha_i = 0$) and the tilted ones ($\alpha_i \neq 0$). For the normal tests, the spin did not change direction and the ball bounced back with a negative deviation angle. For tilted tests, the spin sign did change leading to negative spin ratio and positive deviation angles.

Focusing on the magnitudes, here again a clear distinction could be made between the normal and tilted tests. For normal impacts, the *foam + compact* sample led to a larger deviation angle and a smaller spin ratio than the *foam* specimen. The opposite conclusions prevailed when the incident angle was set equal to 30 or 60°. Finally, the greater the incidence, the larger the difference between the two targets.

A companion study where the incident angle was prescribed to 30° and the incident backspin varied, is presented in Figure 6.

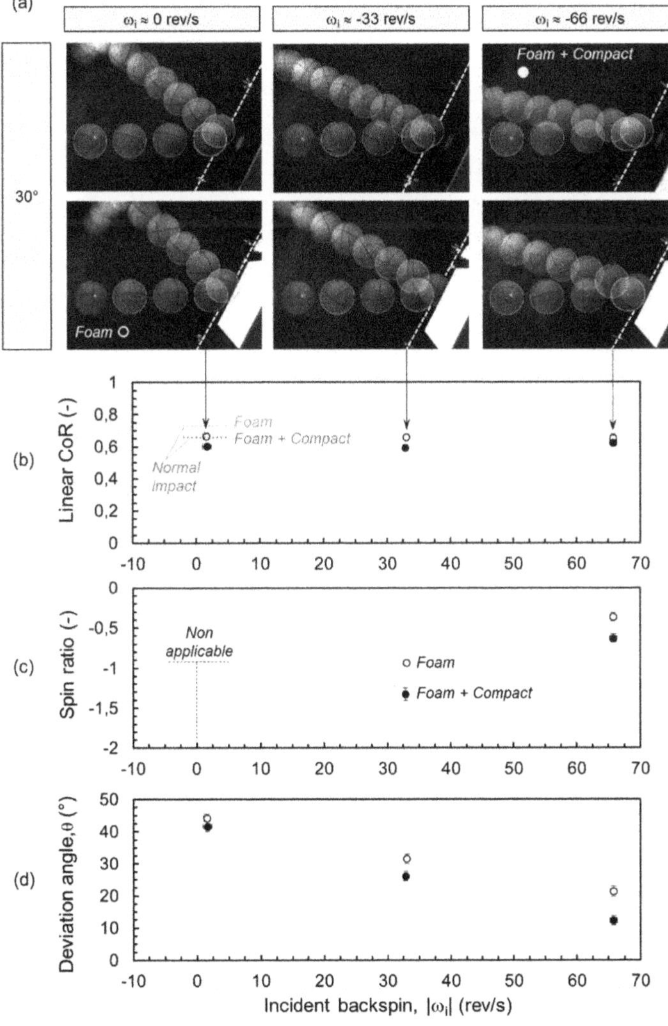

Figure 6. (**a**) Raw images of impact tests performed on the *foam + compact* and *foam* samples with varying angular velocity ($\|\vec{V_i}\| = 14$ m/s, T_{amb}). (**b**) Linear coefficient of Restitution as a function of the incident backspin. (**c**) Spin ratio as a function of the incident backspin. (**d**) Deviation angle as a function of the incident backspin.

Figure 6a displays the raw images of the impact performed on the *foam + compact* (row 1) and *foam* (row 2) samples and Figure 6b–d display respectively the evolution of the linear coefficient of restitution, the spin ratio and the deviation angle as a function of the incident backspin. The linear coefficient of restitution is seen to remain almost constant, with the foam values remaining slightly larger than the foam + compact ones. For the sake of comparisons, the values gathered in [22] for normal impact with no spin were superimposed, evidencing that a deviation from the normal incidence led to a decrease in the CR_L of about 10 percent. Regarding the spin ratio reported in Figure 6b, negative values were found and its magnitudes decreased as the backspin increased, with larger values being found for the *compact + foam* sample. Finally, the deviation angle decreased with increasing backspin. The foam sample exhibited the larger angles compared to the foam + compact sample and the gap increased with increasing backspin.

In summary, distinct discrepancies were experimentally found between the *foam* and *foam + compact* samples response to impact. Whatever the impact condition tested, the linear coefficient of restitution of the *foam + compact* was smaller than the one of the *foam* alone. Also, normal impact aside, impact tests performed on the *foam + compact* target resulted in a larger spin ratio and a smaller deviation angles.

4. Modeling Ball/Pad Impact

As outlined in the introduction, the objective of this section consisted in challenging, with this new set of data, the modelling strategy that had been implemented in [22] to simulate normal impact with no spin. Thus, only minor adjustments were applied to the finite element model. For instance, the constitutive behavior of the constituents remained unchanged. The major modeling assumptions are recalled in the following two first subsections.

4.1. Finite Element Model Description

Explicit 3D finite elements simulations using the commercial software ABAQUS© [26] were pursued to model the impact experiments. With this modeling strategy, the geometry/architecture, the materials intrinsic behavior and the contact properties of the mechanical problem were considered.

The 3D model is presented in Figure 7a. The ball comprises 5058 S4R shell elements with five integration points through the thickness and the polymeric layers is made of 221,598 C3D8R elements: 100,314 elements for the foam layer, 26,028 for the periodic array of perfectly cylindrical pimples and 95,256 elements for the dense top layer of the compact. This distinction has been made since these three layers were modelled as different instances that were bonded together when needed, using the tie function between the shared surfaces as evidenced in the inset of Figure 7b.

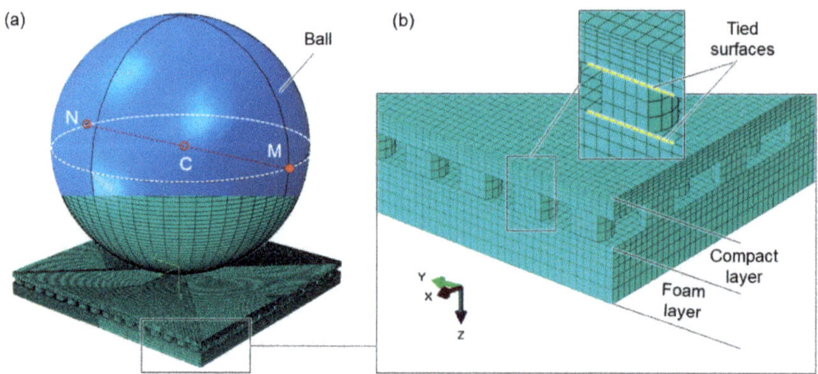

Figure 7. Finite element modeling (FEM). (**a**) 3D model and associated meshes. (**b**) Details on the *foam + compact* modeling.

The nodes of the foam bottom free surface were fixed to replicate its gluing onto the wooden rigid frame. At time t = 0 the ball was positioned 0.3 mm away from the top surface of the target. In order to ensure its rigid motion, the initial velocity was assigned to every node M of the ball and was calculated using the following transport equation:

$$\vec{V}_M = \vec{V}_C + \vec{\omega}_C \wedge \vec{CM},$$ (3)

with $\vec{V}_C$ and $\vec{\omega}_C$ the linear and angular velocities of the ball center C. The velocity was applied for a short duration (10^{-6} s) and the ball was then let free to move (the drag force and gravity were not implemented). Also, constant internal pressure equal to the atmospheric pressure was prescribed as justified elsewhere [8].

The contact between the two bodies was defined *rough* if not otherwise specified. More precisely, the rough condition implies an infinite coefficient of friction (total adhesion), suggesting that all relative sliding motion between two contacting surfaces is prevented [27]. It is worth mentioning that this hypothesis was used in [22] to successfully predict the linear coefficient of restitution of the various polymeric targets normally impacted. This hypothesis is strongly challenged and discussed with this new set of data.

4.2. Intrinsic Behavior of the Constituents

A rate-dependent dissipative elastic description was used to describe the ball parent glassy material since it did not overcome its elastic limit under the large set of impact tested. The foam and the compact were defined as rate-dependent dissipative hyperelastic materials. The same constitutive equation was used for both layers as justified in [22].

Dynamic mechanical analysis (DMA) tests and time-temperature superposition principles were used to characterize the dissipative rate dependence elastic behavior of the different materials. Additional quasi-static compression test was performed on the foam system to characterize its large non-linear deformation behavior. An overview of the DMA response in the tensile mode of the ball parent material and the foam are displayed in Figure 8a,b respectively, and the large strain compression response of the foam is presented in Figure 8c. The reader should refer to [22] for further details on the methods and testing conditions.

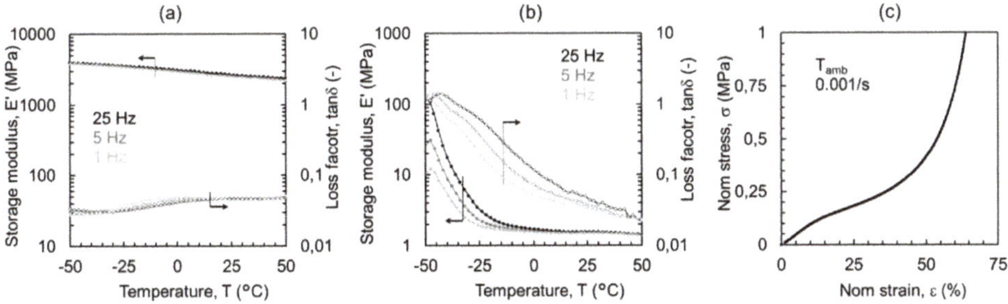

Figure 8. (**a**) Isochronal dynamic mechanical analysis (DMA) profiles of the ball parent material in the [−50; 50] °C temperature range. (**b**) Isochronal DMA profiles of the foam material in the [−50; 50] °C temperature range. (**c**) Compression stress strain response of the foam parent material. N.B. This figure is a reproduction of Figure 3 in [22].

In ABAQUS©, Prony series fitting the experimentally based time-dependent shear modulus of visco-elastic materials G(t) were implemented. Nine discrete Maxwell elements were identified for the ball and the foam (and compact). Constant Poisson ratio were allocated to the polymeric constituents to control the volumetric behavior. Finally, table data of the experimental strain-stress response was used to fit the isotropic Ogden strain energy potential for the long-time hyperelastic behavior of the polymeric coatings layers [27]. It is worth noting that the Mullins and Payne effects of the elastomeric layers were neglected and their potential effects minimized experimentally (by testing and characterizing fatigued samples on the one hand and controlling the recovering time between consecutive impact tests) [28].

Also, the foam was modeled as a continuum media whereas the exact architecture of the compact layer was modeled. Consequently, their density and Poisson ratio differ. Almost incompressible Poisson ratio ($^C v = 0.49$) and density $^C \rho = 1000$ kg/m^3 were assigned to the compact parent material whereas effective values of $^F \rho = 490$ kg/m^3 and $^F v = 0.41$ (determined using video-extensometry) were measured and implemented for the foam.

4.3. Experiments/Simulations Comparisons

While the ball's simulated incident trajectory was an input, its reflected trajectory needed to be determined in order to compute the aforementioned metrics. Thus, the position of the center of the ball led to the determination of deviation angle and the linear coefficient of restitution and was calculated from the coordinates at any time increment of two nodes, denoted M and N in Figure 7a, located away from the contact region. The center was half way between M and N, assuming that these two nodes corresponded at any time to the ball's diameter.

With this assumption, the rotation angle β of the ball with respect to the out of plane *x-axis* was determined as follows:

$$\beta = \mathrm{atan}\left(\frac{z_N - z_M}{y_N - y_M}\right), \tag{4}$$

with $\{y_N, z_N\}$ and $\{y_M, z_M\}$ the in-plane y- and z-coordinates of the N and M points respectively. The associated angular velocity (spin) was then determined from two consecutive time increments:

$$\omega = \frac{\Delta\beta}{\Delta t}, \tag{5}$$

with $\Delta\beta$ and Δt the angle and time increments, respectively.

Figure 9 presents the simulated vertical displacement mapping for a ball impacting the *foam + compact* target at a 60° angle, 14 m/s and −66 rev/s. Four screenshots are displayed corresponding to the initial time (t_0), first contact (t_1), contact at maximum depth (t_2) and end time (t_{end}). A close up view of the through-thickness deformation in the contact region at time t_2 is added to evidence the complex deformation pattern of the target layers. It is worth noting that no buckling of the ball occurred in this case and that the pimples were significantly sheared.

The positions of the N and M equatorial points, and a fictitious line connecting them are also displayed. As discussed above, their time positions were necessary for the metrics calculations. Also, the trajectories of the M, N and calculated C points are plotted in Figure 10. The positions for the times selected in Figure 9 are presented and the spin inversion is clearly evidenced.

In the previous study focusing on the normal impact with no spin, the linear coefficient of restitution was the sole relevant metric monitored. Also, finite element simulations with rough contact conditions resulted in fair predictions of the experimental data compared to simulations with frictionless ones. A similar study had been conducted in this work for the *foam* sample impacted under various incidences and spins. The results are presented in Figure 11 where the experimental and simulated linear coefficient of restitutions are compared.

Figure 11a focuses on impacts with fixed backspin (−66 rev/s) and varying incidence and Figure 11b presents the results for impacts at constant incidence (30°) and varying backspin. Fair predictions using rough contact conditions were obtained whereas frictionless simulations clearly over predicted the experimental results. With this broad set of impact conditions, the numerical predictions capability using rough contact conditions could be further challenged when tracking the two additional metrics that had been defined.

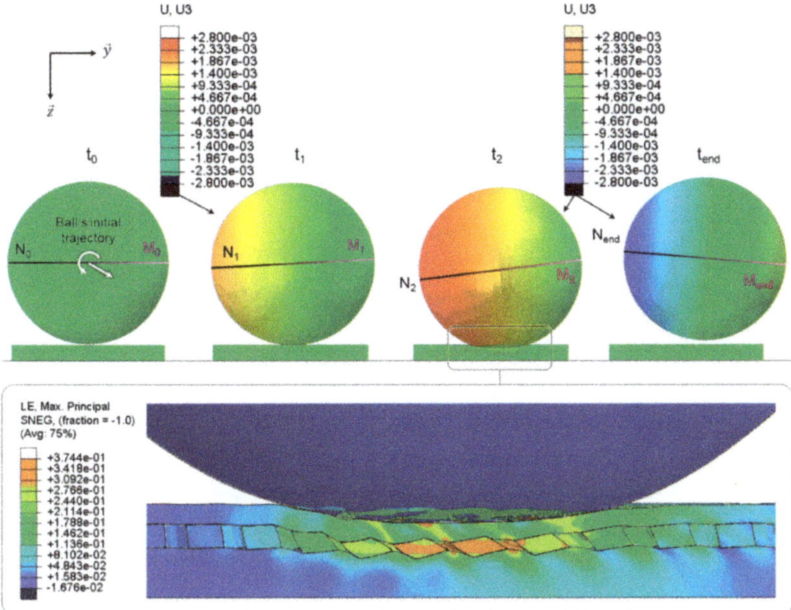

Figure 9. Simulated vertical displacement mapping of a ball impacting the *foam + compact* target at 4 distinct times ($\alpha_i = 60°$ angle, $V_i = 14$ m/s and $\omega_i = -66$ rev/s). The inset figure focuses on the through-thickness deformation profile at maximum depth during contact.

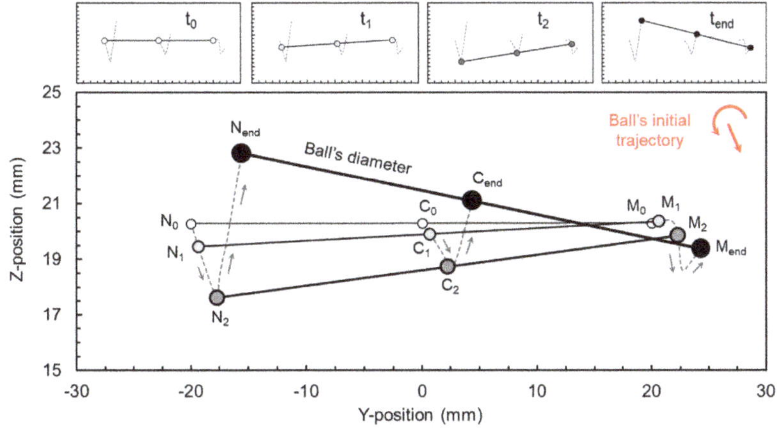

Figure 10. In-plane trajectories of the M, N and calculated C points of the ball. ($\alpha_i = 60°$ angle, $V_i = 14$ m/s and $\omega_i = -66$ rev/s). N.B. different scales were used for y and z.

The experiment/simulation comparisons are presented in Figure 12 for impact tests at constant velocities ($V_i = 14$ m/s and $\omega_i = -66$ rev/s) and varying incidence, performed on the *foam* (Figure 12a) and *foam + compact* (Figure 12b) targets. For both the *foam* and the *foam + compact* targets, the magnitude of the linear coefficient of restitution and its decrease with increasing the incident angle were fairly predicted (top figures in Figure 12).

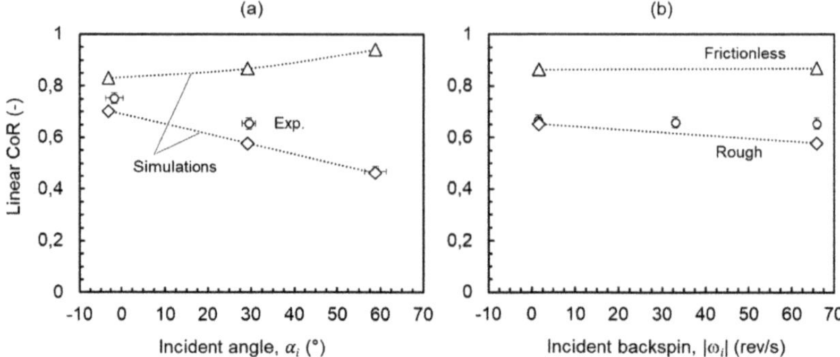

Figure 11. Effect of the ball/foam contact properties on the predicted linear coefficient of restitutions. (a) Foam impacted at varying incidence and constant velocities ($V_i = 14$ m/s and $\omega_i = -66$ rev/s). (b) Foam impacted at varying incident backspin and constant angle ($\alpha_i = 30°$) and linear velocity ($V_i = 14$ m/s).

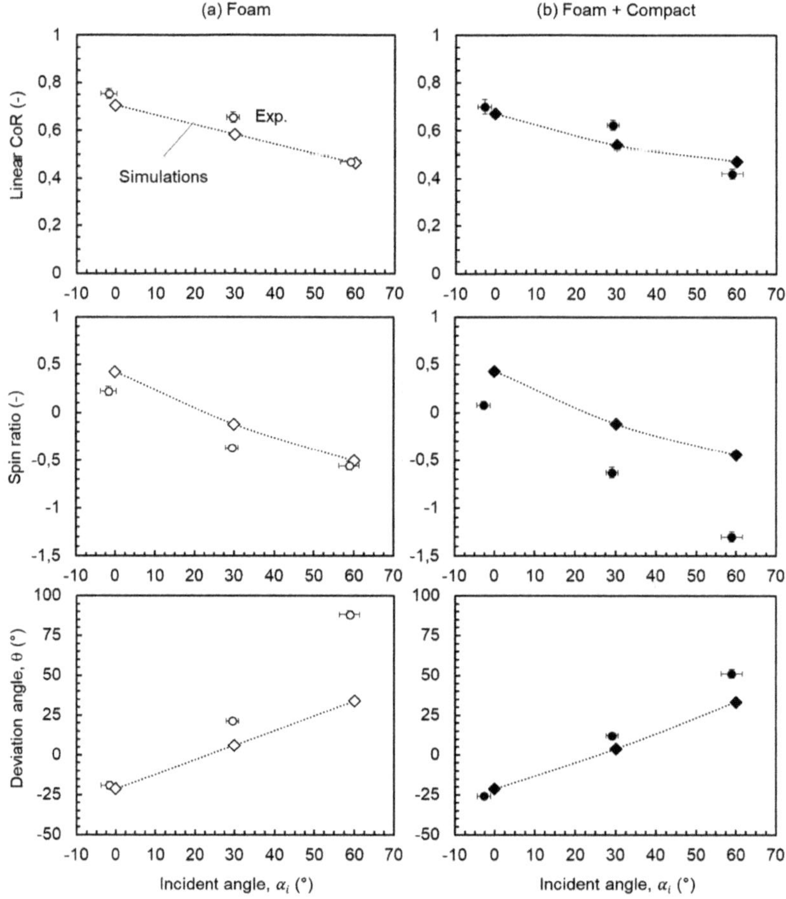

Figure 12. Model/experiments confrontations of the three metrics of performance as a function of the incident angle for impact ($V_i = 14$ m/s and $\omega_i = -66$ rev/s). (a) *foam* sample. (b) *foam + compact* sample.

Furthermore, the qualitative evolutions of the two other metrics, that were an inversion and then an increase of the absolute value of the spin ratio and an increase of the deviation angle, were both captured. Yet, the experimental and simulated magnitudes strongly differed. More precisely, limited differences could be observed between the simulated metrics of the two targets whereas large experimental discrepancies were already reported: normal impact aside, impacts on the *foam + compact* sample resulted in larger spin ratio magnitudes and smaller deviation angles θ.

These differences suggested that the ball/foam and ball/compact contacts were different. At first glance, the ball/foam friction coefficient seemed smaller than the compact/foam one. Nevertheless, the potential difference between the deforming contact geometries could not be ignored. Thus, complementary studies of the contact shape with account for the refined and potentially different contact properties between the foam and the compact layers need to be pursued. Numerically, parametric study with varying coefficient of friction (CoF) can be performed. Experimentally, its measurement can be obtained as reported in [29,30] for instance. These refinements are further justified since the reversible buckling of the ball (leading to very different contact areas) only occurred for certain experimental conditions tested. These investigations are currently pursued. Also, in the present study, the contact forces (normal and tangent) were not considered. Future works will consider the implementation of a force plate to measure the impact of normal and tangent forces and compare with the ones predicted by the finite element analysis, as in [18,19]. This will probably help in tuning the friction coefficient between the ball and the covering.

5. Conclusions

The impact of a table tennis ball onto part or the complete polymeric racket layers had been investigated for various incident angles and spins. Three metrics of performance, namely the linear coefficient of restitution CR_L, the spin ratio SR and the deviation angles θ were monitored evidencing strong differences between the *foam* and *foam + compact* layers. In short, normal spin aside, the ball impacting the *foam* pad exhibited smaller reflected spin and larger reflected angle than the one impacting the *foam + compact*.

The numerical prediction of the impact had been attempted using a previously implemented 3D finite element model that happened to successfully predict the linear coefficient of restitution of the same target under normal impact with no spin. The rate dependent dissipative constitutive equations of the ball and polymeric layers were determined based on DMA and the large strain hyperelastic response was obtained from quasi-static compression tests.

While the linear coefficient of restitution was fairly predicted, the two other metrics failed to be successfully simulated, revealing the key role of the contact condition. The simplistic rough contact condition needs to be refined.

However, the numerical tool comprises all the required elements to study the impact of a table tennis ball on the racket polymeric layers. Ongoing work focuses on a better account for the contact, and detailed analysis of the simulated contact region is attempted in order to get a better sense of the role of the contact geometry. In addition, the in-depth study of the energy balance (strain, kinetic, viscous, friction components) is currently being pursued to identify the key mechanisms driving the overall performance of the polymeric layers impacted by a table tennis ball.

Author Contributions: Conceptualization, R.G.R., L.M., S.M. and N.H.; methodology, R.G.R., L.M., S.M. and N.H.; software, R.G.R. and L.M.; validation, R.G.R., L.M., S.M. and N.H.; formal analysis, R.G.R. and L.M.; investigation, R.G.R. and L.M.; resources, R.G.R., L.M., S.M. and N.H.; data curation, R.G.R., L.M.; writing—original draft preparation, R.G.R., L.M.; writing—review and editing, R.G.R., L.M..; visualization, R.G.R.; supervision, R.G.R., L.M.; project administration, N.H.; funding acquisition, N.H.

Funding: This research received no external funding.

Acknowledgments: The long-term collaboration between INSA Lyon and the French company Cornilleau© consists of incremental student projects over the years. Consequently, the numerous students who have been involved over the years deserved to be acknowledged for their implications and efforts. Here they are, chronologically listed: T. Huin, A. Drillon, C. Bonnard, Z. Hui, L. Marin Curtoud, D. Quanshangze. Special thanks also go to H. Lourenco for his help during the experimental campaigns performed at Cornilleau in the early years of this collaboration.

Conflicts of Interest: The authors declare no conflict of interest.

References

1. Kamijima, K.; Ushiyama, Y.; Yasaka, T.; Ooba, M. Effect of different playing surfaces of the table on ball bounces in table tennis. *Int. J. Table Tennis Sci.* **2013**, *1*, 11–14.
2. Nagurka, M.; Huang, S. A mass-spring-damper model of a bouncing ball. In Proceedings of the 2004 American Control Conference, Boston, MA, USA, 30 June–2 July 2004.
3. Nakashima, A.; Ito, D.; Hayakawa, Y. An online trajectory planning of struck ball with spin by table tennis robot. In Proceedings of the 2014 IEEE/ASME International Conference Advanced Intelligent Mechatronics (AIM), Besançon, France, 8–11 July 2014.
4. Nakashima, A.; Okamoto, T.; Hayakawa, Y. An online estimation of rotational velocity of flying ball via aerodynamics. *IFAC Proc. Vol.* **2014**, *47*, 7176–7181. [CrossRef]
5. Nonomura, J.; Nakashima, A.; Hayakawa, Y. Analysis of effects of rebounds and aerodynamics for trajectory of table tennis ball. In Proceedings of the SICE Annual Conference 2010, Taipei, Taiwan, 18–21 August 2010.
6. An, C.-C.; Hsu, H.Y.; Sun, Y.T.; Ke, L.D.; Hsu, T.G.; Ting, C.C. Developing an audio analyzer for instantaneous stroke position identification on table tennis racket to assist technical training. *Measurement* **2018**, *115*, 73–79. [CrossRef]
7. Widenhorn, R. The physics of juggling a spinning ping-pong ball. *Am. J. Phys.* **2016**, *84*, 936–942. [CrossRef]
8. Zhang, X.; Tao, Z.; Zhang, Q. Dynamic behaviors of visco-elastic thin-walled spherical shells impact onto a rigid plate. *Lat. Am. J. Solids Struct.* **2014**, *11*, 2607–2623. [CrossRef]
9. Dong, X.; Gao, Z.; Yu, T. Dynamic crushing of thin-walled spheres: An experimental study. *Int. J. Impact Eng.* **2008**, *35*, 717–726. [CrossRef]
10. Pauchard, L.; Rica, S. Contact and compression of elastic spherical shells: The physics of a 'ping-pong' ball. *Philos. Mag. B* **1998**, *78*, 225–233. [CrossRef]
11. Ruan, H.; Gao, Z.; Yu, T. Crushing of thin-walled spheres and sphere arrays. *Int. J. Mech. Sci.* **2006**, *48*, 117–133. [CrossRef]
12. Shorter, R.; Smith, J.; Coveney, V.; Busfield, J. Axial compression of hollow elastic spheres. *J. Mech. Mater. Struct.* **2010**, *5*, 693–705. [CrossRef]
13. Cho, H.; Rinaldi, R.G.; Boyce, M.C. Constitutive modeling of the rate-dependent resilient and dissipative large deformation behavior of a segmented copolymer polyurea. *Soft Matter* **2013**, *9*, 6319–6330. [CrossRef]
14. Rinaldi, R.G.; Hsieh, A.; Boyce, M. Tunable microstructures and mechanical deformation in transparent poly (urethane urea) s. *J. Polym. Sci. Part B Polym. Phys.* **2011**, *49*, 123–135. [CrossRef]
15. Tiefenbacher, K.; Durey, A. The impact of the Table Tennis Ball on the Racket (Backside Coverings). *Int. J. Table Tennis Sci.* **1994**, *2*, 1–14.
16. Tiefenbacher, K.; Seydel, R.; Durey, A. Analysis of the influence of special equipment materials on decisive strokes. *Int. J. Table Tennis Sci.* **1996**, *2*, 51–60.
17. Cross, R. Grip-slip behaviour of a boucing ball. *Am. J. Phys.* **2002**, *70*, 1093–1102. [CrossRef]
18. Goodwill, S.R.; Kirk, R.; Haake, S.J. Experimental and finite element analysis of a tennis ball impact on a rigid surface. *Sports Eng.* **2005**, *8*, 145–158. [CrossRef]
19. Allen, T.; Goodwill, S.; Haake, S. Experimental validation of a tennis ball finite-element model. In *Tennis Science and Technology 3*; Miller, S., Capel-Davies, J., Eds.; International Tennis Federation: London, UK, 2007; pp. 23–30.
20. Allen, T.; Goodwill, S.; Haake, S. Experimental validation of a tennis ball finite-element model for different temperatures (P22). In *The Engineering of Sport 7*; Springer: Paris, France, 2009.
21. Allen, T.; Haake, S.; Goodwill, S. Comparison of a finite element model of a tennis racket to experimental data. *Sports Eng.* **2009**, *12*, 87–98. [CrossRef]
22. Rinaldi, R.G.; Manin, L.; Bonnard, C.; Drillon, A.; Lourenco, H.; Havard, N. Non linearity of the ball/rubber impact in table tennis: Experiments and modeling. *Procedia Eng.* **2016**, *147*, 348–353. [CrossRef]
23. Carre, M.J.; Haake, S.; Baker, S.W.; Newell, A. The analysis of cricket ball impacts using digital stroboscopic photography. In *The Engineering of Sport—Design and Development*; Blackwell: Oxford, UK, 1998; pp. 379–386.

24. Theobalt, C.; Albrecht, I.; Haber, J.; Magnor, M.; Seidel, H.P. Pitching a baseball: tracking high-speed motion with multi-exposure images. *ACM Trans. Graph.* **2004**, *23*, 540–547. [CrossRef]

25. Allen, T.; Ibbitson, J.; Haake, S. Spin generation during an oblique impact of a compliant ball on a non-compliant surface. *Proc. Inst. Mech. Eng. Part P J. Sports Eng. Technol.* **2012**, *226*, 86–95. [CrossRef]

26. *ABAQUS, Analysis User's Manual, Version 6.12*; Karlsson & Sorensen, Inc.: Pawtucket, RI, USA, 2012.

27. *ABAQUS: Theory manual*; Hibbit, Karlsson & Sorensen, Inc.: Pawtucket, RI, USA, 1997.

28. Ward, I.M.; Sweeney, J. *Mechanical Properties of Solid Polymers*; John Wiley & Sons: Hoboken, NJ, USA, 2012.

29. Varenberg, M.; Varenberg, A. Table tennis rubber: Tribological characterization. *Tribol. Lett.* **2012**, *47*, 51–56. [CrossRef]

30. Varenberg, M.; Varenberg, A. Table Tennis: Preliminary Displacement in Pimples-Out Rubber. *Tribol. Lett.* **2014**, *53*, 101–105. [CrossRef]

 *applied sciences*

Article

Characterization of Maple and Ash Material Properties for the Finite Element Modeling of Wood Baseball Bats

Joshua Fortin-Smith [1], James Sherwood [1,*], Patrick Drane [1] and David Kretschmann [2]

[1] Baseball Research Center, Mechanical Engineering, University of Massachusetts Lowell, 1 University Avenue, Lowell, MA 01854, USA; Joshua_FortinSmith@student.uml.edu (J.F.-S.); Patrick_Drane@uml.edu (P.D.)

[2] U.S. Forest Products Laboratory, US Forest Service, 1 Gifford Pinchot Drive, Madison, WI 53726, USA; dekretschmann@gmail.com

* Correspondence: James_Sherwood@uml.edu; Tel.: +1-978-934-3313

Received: 1 May 2018; Accepted: 7 November 2018; Published: 15 November 2018

Abstract: To assist in developing a database of wood material properties for the finite element modeling of wood baseball bats, Charpy impact testing at strain rates comparable to those that a wood bat experiences during a bat/ball collision is completed to characterize the failure energy and strain-to-failure as a function of density and slope-of-grain (SoG) for northern white ash (Fraxinus americana) and sugar maple (Acer saccharum). Un-notched Charpy test specimens made from billets of ash and maple that span the range of densities and SoGs that are approved for making professional baseball bats are impacted on either the edge grain or face grain. High-speed video is used to capture each test event and image analysis techniques are used to determine the strain-to-failure for each test. Strain-to-failure as a function of density relations are derived and these relations are used to calculate inputs to the *MAT_WOOD (Material Model 143) and *MAT_EROSION material options in LS-DYNA for the subsequent finite element modeling of the ash and maple Charpy Impact tests and for a maple bat/ball impact. The Charpy test data show that the strain-to-failure increases with increasing density for maple but the strain-to-failure remains essentially constant over the range of densities considered in this study for ash. The flat response of the ash data suggests that ash-bat durability is less sensitive to wood density than maple-bat durability. The available SoG results suggest that density has a greater effect on the impact failure properties of the wood than SoG. However, once the wood begins to fracture, SoG plays a large role in the direction of crack propagation of the wood, thereby determining if the shape of the pieces breaking away from the bat are fairly blunt or spear-like. The finite element modeling results for the Charpy and bat/ball impacts show good correlation with the experimental data.

Keywords: baseball; bat; Charpy; finite element; impact; wood

---

1. Introduction

In 2008, Major League Baseball (MLB) commissioned a team of experts comprised of wood scientists and bat performance test engineers to investigate options for reducing the rate of bats breaking during games. While there were no data to support if there was or was not an increasing rate of bats breaking into multiple pieces during games, there was a perception of an increased breaking rate by the league and by fans. The perceived rise in bat breakage rates coincided with the increase in popularity of the maple wood species as a bat preference amongst players in MLB. Historically, the ash wood species was the wood of choice for players in MLB up until the entry of maple bats in the game in the late 1990s. Maple became the trendy choice after Barry Bonds' historic 2001 season in which he set a MLB single-season record of 73 homeruns using a maple bat.

To develop a baseline on the rate of bat breakage, MLB completed a two-and-a-half-month collection of broken bats in August 2008. The lessons learned from this and subsequent wood bat collections led to new requirements in the WBBS (Wooden Baseball Bat Specifications), specifically ash bats continue to be impacted on the edge grain, maple bats change to be impacted on the face grain and the slope-of-grain (SoG) of the wood used to manufacture baseball bats to be within $\pm 3°$ of the centerline of the bat longitudinal axis.

To complement the wood bat collections, research efforts were conducted to investigate how finite element modeling can aid in explaining how bat profile and wood quality relate to bat durability [1]. Previous research studies conducted on metal, composite and wood baseball bats utilizing the finite element method have been demonstrated to be an effective way of analyzing the effect of high-speed impact on the bat after impact with the baseball occurs [2–5]. These finite element models have been found to be valuable tools for providing insight into the mechanical response of wood baseball bats over a range of impact speeds [6]. However, the description of the material properties to date has been limited to what can be extracted from quasistatic four-point bend testing of dowels and from the Wood Handbook [7]. These quasistatic test programs were completed at the USDA (United States Department of Agriculture) Forest Product Labs (FPL) and have documented the Modulus of Elasticity (MOE) and Modulus of Rupture (MOR) for ash, maple and yellow birch as a function of wood density [8]. To further improve the level of correlation between these finite element models and lab-simulated bat/ball impacts, the material behavior for these wood species must also be characterized at strain rates comparable to those experienced during game-condition bat/ball collisions.

In the current research, Charpy Impact testing of ash and maple was completed to characterize the high strain-rate behavior of these wood species. To date the documented research on the dynamic behavior of wood during impact is fairly limited [9,10] compared to other engineering materials. By following the testing methodologies that were used by the Federal Highway Administration (FHWA) [11] to characterize southern yellow pine and applying them to maple and ash, the current high-speed testing characterized the material behavior of these wood species further than is currently available. Once characterized, the resulting material properties can be prescribed in the *MAT_WOOD (Material Model 143) and *MAT_EROSION material models in LS-DYNA for use in finite element analyses of bat/ball impacts. In addition to the mechanical testing data from the FPL and the Charpy programs, past test data for maple and ash bats in the ADC Bat Durability Testing System that are available at the UMass Lowell Baseball Research Center (UMLBRC) [1,12] along with properties taken from the Wood Handbook [7] were utilized to further refine wood properties. With a level of confidence in the material parameter inputs established, the finite element modeling can be used to support scientifically defensible arguments for proposing future changes to wood bat specifications aimed at reducing bat breakage rates. In some cases, these changes can be essentially transparent to players as to what they have come to expect in bat feel and performance.

This paper describes the Charpy test program that was used to characterize the strain-to-failure of ash and maple wood specimens under dynamic loading and the subsequent finite element modeling using the material parameters derived from the test program. Finite element modeling of the Charpy tests is performed to show that the models are capable of replicating the tests from which the material parameters were derived. With a good level of comfort established for the ability of the finite element models to simulate a Charpy impact tests, the modeling approach is shown to be applicable to simulating the mechanical behavior of wood bats to ball impacts.

2. Materials and Methods

Quasistatic and dynamic material testing was conducted to characterize the material behavior of maple and ash. Four-point bend testing of wood dowels was conducted at the FPL in 2009 to quantify the relationship between wood density and the MOE (Modulus of Elasticity) and wood density and the MOR (Modulus of Rupture). In the current research, Charpy Impact testing was conducted to characterize wood behavior during dynamic impact to investigate the relationship between wood density and strain-to-failure.

2.1. Existing Test Data

In 2009, quasistatic four-point bend testing of ash and maple dowels was completed at the FPL in Madison, WI [8]. These test data quantified how the MOE and MOR vary as a function of density for ash and maple. The density range tested for each wood species corresponded to the range of densities that are used to manufacture MLB-quality baseball bats. Figures 1 and 2 show MOE-density and MOR-density results, respectively, for the ash and maple dowels. These dowels had slope-of-grain (SoG) angles between ±3°, which is the current range of SoGs that are allowed for the manufacture of the baseball bats used by major-league players. A 0° SoG is when the direction of the wood grain is aligned with the axis of the bat. Figures 1 and 2 show how the MOE and MOR, respectively, of each of the wood species increases with increasing density. These test data reflect the wide scatter in material properties that are typical for wood. A linear regression on each set of data was completed. The resulting equations are:

$$Maple\ MOE\ (Msi) = 39.96 \times \rho + 1.279 \tag{1}$$

$$Maple\ MOR\ (psi) = 1002663 \times \rho - 2554 \tag{2}$$

$$Ash\ MOE\ (Msi) = 122.5 \times \rho - 0.875 \tag{3}$$

$$Ash\ MOR\ (psi) = 1383652 \times \rho - 10117 \tag{4}$$

where ρ is the density in lb/in^3.

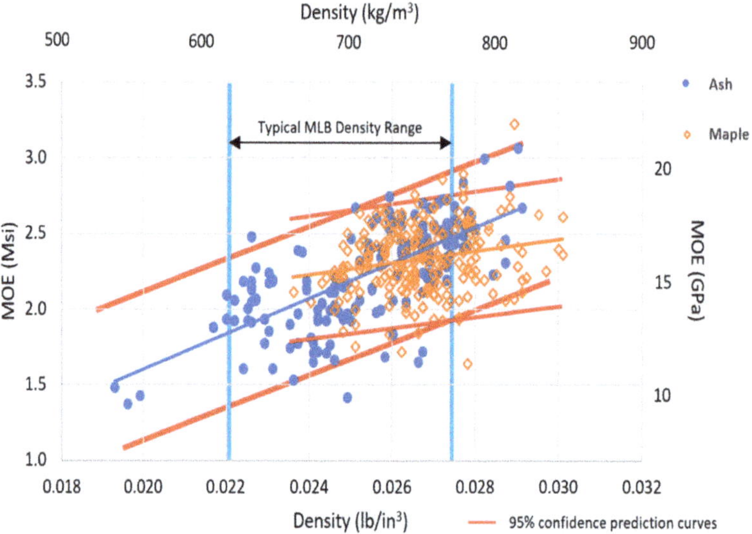

Figure 1. Modulus of Elasticity (MOE) test results for ash and maple dowels with slope-of-grain angles spanning the range of ±3°.

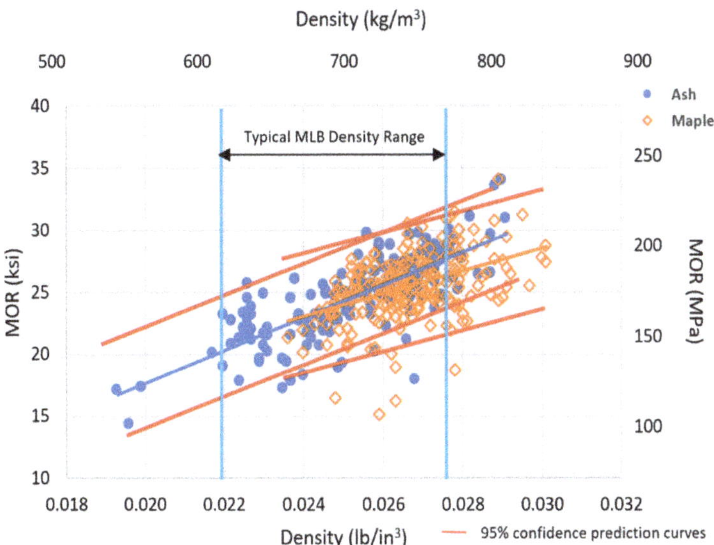

Figure 2. Modulus of Rupture (MOR) test results for ash and maple dowels with slope-of-grain angles spanning the range of ±3°.

2.2. Dynamic High-Rate Material Testing

Wood is a viscoelastic material and as a result, its strain-to-failure is a function of strain rate. For quantifying the strain-to-failure as a function of strain rate, the Charpy Impact test was used for all dynamic wood testing. The Charpy Impact is a test that can be conducted at the strain rates that are representative of what ash and maple experience during bat/ball impacts. The range of strain rates that occur during a collision have not been measured directly but an estimate of the strain rates ranging from 10^1–10^4 s^{-1} during such a collision have been concluded from what has been seen in preliminary finite element analyses of bat/ball impacts using LS-DYNA.

A standard Charpy test involves the use of a pendulum swinging a hammer to impact a specimen of known geometry [13]. The height that the pendulum reaches at the end of the swing following the breaking of the test sample is recorded and this height is used to determine the energy that was required to break the test specimen. This energy is the only direct output of the Charpy test. Thus, additional complementary analyses must be completed to quantify the strain rate that was experienced by the test specimen. A Redlake HG100k high-speed camera in conjunction with Motion Studio camera software was set up to record each sample's impact and subsequent breakage using a frame rate of 30,000 frames/s at a resolution of 160 pixels wide by 168 pixels high. The captured high-speed video facilitated the complementary analysis to quantify the strain rate experienced by the wood specimens during each test event.

2.3. Sample Preparation

Charpy test specimens were made from blocks of ash and maple that spanned the range of densities from 0.022–0.029 lb/in^3 (608.9–747.4 kg/m^3). This range is representative of the span of wood densities that are approved for making professional baseball bats. Wood samples were cut to a square cross-section geometry within the dimensions as specified by ASTM Standard D6110-10 Standard Test Method for Determining the Charpy Impact Resistance of Notched Specimens of Plastics, that is, $5.0 \times 0.5 \times 0.5$ in. ($12.7 \times 1.27 \times 1.27$ cm) (length × width × thickness) [13]. For this testing, samples were un-notched for the purpose of determining the strain required to initiate fracture of the solid wood geometry, which is similar to a baseball bat in that a bat would not be used if cracked or notched, during a standard Charpy test. Utilizing ImageJ software, the edge (radial) and face (tangential) SoGs were measured for each sample [14]. Figure 3 shows an example of the grain line used to measure SoG on the samples within ~0.2°. Samples were conditioned for a minimum of 14 days in 50 ± 10% relative humidity and 72 ± 2 °F (22 ± 1 °C) before doing any testing. This same conditioning is what is used before testing wood baseball bats at the UMLBRC.

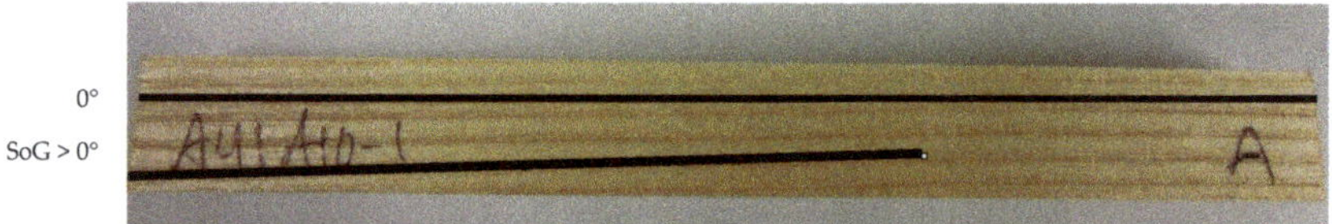

Figure 3. Example of a radial slope-of-grain measurement.

2.4. Test Procedure

The Charpy Impact test measures the resistance of a material to breakage by the flexural shock that is induced by a pendulum of specified weight to break a test specimen with a single swing [13]. The output of the test is the energy required to break a specimen of a prescribed size. Samples were impacted with a 10.0 ft-lb (13.6 N-m) hammer on the edge grain and face grain (1) to investigate the differences between edge-grain versus face-grain impacts within each wood species and (2) to conclude a strain-to-failure as a function of wood density for each wood species.

The information collected from the Charpy test included (1) the energy required to break the sample and (2) the high-speed video. The break energy of each sample was directly read from the dial of the Charpy impact tester. The break energy indicates the loss of energy of the pendulum to fracture the sample. ImageJ was used to examine the high-speed video and to quantify the maximum deflection of the sample just before breaking. The process for measuring this maximum deflection was first to identify the position of a point on the non-impact side of the sample before impact and then to locate the same point directly before failure occurs. An example of this process is presented in Figure 4 with the image on the left being the starting position of the point and the image on the right showing the same point directly before failure occurs. It should be noted that this method is imperfect and some degree of measurement error could be included within the individual measurements, thus the average result of the entire sample set and one standard deviation is reported. Using the standard deflection equation for three-point bending in beam theory, the strain-to-failure can be estimated using Equation (5),

$$\varepsilon = \frac{12y\delta}{L^2} \tag{5}$$

where L is the effective length of specimen between the supports, δ is the max deflection and y is the half-thickness.

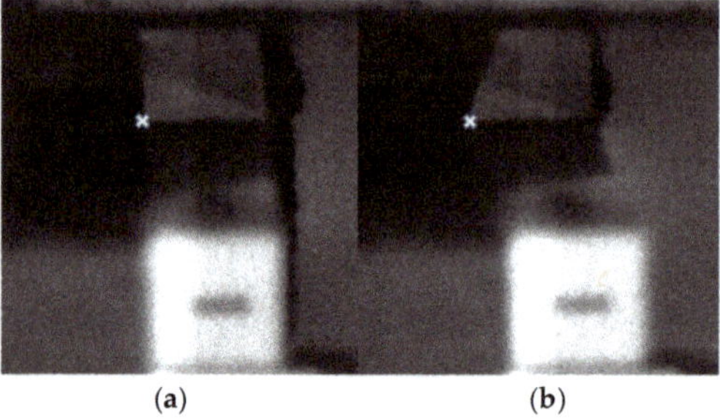

Figure 4. Maximum Deflection Measurement (**a**) just before impact and (**b**) just before fracture. The white "✖" denotes the tracked point before impact and at the initiation of fracture.

3. Results

The overall mean of the results of the Charpy testing is presented in Table 1. The results show that maple exhibited a lower failure energy, a lower maximum-deflection and a lower strain-to-failure for both edge- and face-grain loadings relative to ash. Maple exhibited a higher failure energy when loaded in the face-grain direction compared to edge-grain loading. This higher failure energy for face-grain impacts was expected because maple is known to exhibit better durability (higher relative bat/ball impact breaking speed) when impacted on the face grain compared to bat/ball impacts on the edge grain [15]. Ash performed better overall than maple with a greater deflection at breaking and hence, a greater strain-to-failure when impacted on the edge grain, than maple when impacted on either the edge or the face grain. Ash exhibited superior failure energy when impacted on the face grain. From a strain-to-failure perspective, the ash samples performed better when impacted in edge-grain loading, which is consistent with the preferred impact surface of ash bats due to flaking of the wood cells inherent to repeated face-grain impacts.

Table 1. Summary of mean values with one standard deviation for Charpy Impact testing of Major League Baseball (MLB)-quality Ash and Maple.

Species	Impact Surface	Number of Samples	Strain to Failure	Max Deflection in. (cm)	Energy ft-lbf (N-m)
Ash	edge	23	0.0265 ± 0.002	0.148 ± 0.0122 (0.376 ± 0.0310)	6.48 ± 1.02 (8.79 ± 1.38)
Ash	face	12	0.0259 ± 0.003	0.146 ± 0.0139 (0.371 ± 0.0353)	7.19 ± 1.14 (9.75 ± 1.55)
Maple	edge	21	0.0236 ± 0.002	0.133 ± 0.0134 (0.338 ± 0.0340)	6.30 ± 1.33 (8.54 ± 1.80)
Maple	face	33	0.0250 ± 0.004	0.140 ± 0.0191 (0.356 ± 0.0485)	6.41 ± 1.48 (8.69 ± 2.01)

3.1. Density

Figure 5 shows the strains-to-failure for maple samples impacted on the face grain and for ash samples impacted on the edge grain, respectively and Figure 6 shows the failure energy for the same wood species and impact-surfaces, respectively. These two combinations of wood species and impact-surface type are of primary interest because these are the same combinations that are currently prescribed for on-field bat/ball impacts. There is wide scatter among the data points in each of these four figures but this type of scatter is not unusual in wood testing. Despite the wide scatter, some trends and differences can be noted amongst the data for each of the wood species as a function density.

Linear regressions (as denoted by the solid and dashed black lines) of the Charpy data are included in Figure 5. For the maple sample set (Figure 5, open data points), the strain-to-failure exhibits a much steeper response with respect to density than the slope of the ash data (Figure 5, solid data points). The flatter slope of the ash data suggests that the durability of ash bats is less sensitive to wood density than maple bats. As used in this paper, bat durability is defined as the relative bat/ball impact speed at specific axial location along the bat—the greater the relative bat/ball impact speed, the greater the bat durability [15]. On this basis and in terms of baseball bats, it is important when using a maple wood to maximize the wood density to achieve the best impact properties for the desired profile. For ash bats, the density, while still a concern, is not as essential to the impact properties as the impact face is, with edge grain impact being the preferred impact surface. An analogous interpretation of the failure energy plot (Figure 6) can be made.

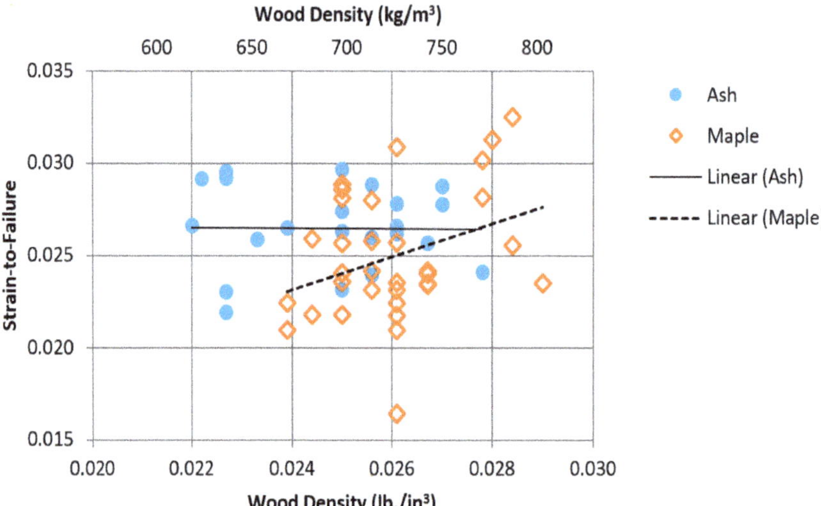

Figure 5. Strain-to-failure as a function of wood density for ash samples impacted edge-grain and maple samples impacted face-grain.

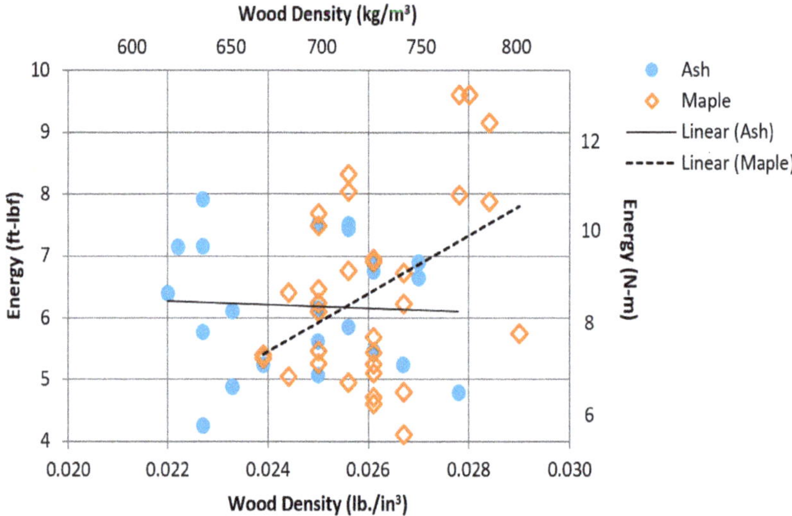

Figure 6. Failure energy as a function of wood density for ash samples impacted edge-grain and maple samples impacted face-grain.

3.2. Slope of Grain

The Charpy data were sorted into three SoG groupings, specifically (1) |SoG| < 1°, (2) 1° < |SoG| < 3° and (3) |SoG| > 3°. Absolute values are used in the SoG grouping due to the symmetry involved in the Charpy Impact test. Because it is well-known that the stress-to-failure decreases with increasing absolute value of the SoG [16,17], it was anticipated that the test data would show a similar trend for the strain-to-failure as a function of SoG for maple and for ash.

Figure 7 is the same plot as Figure 5 where the ash and maple data have been sorted into three SoG groupings. It is challenging to see a clear decrease in the strain-to-failure with increasing SoG in Figure 7. However, upon review of the mean strain-to-failure and standard deviation of the data set in Table 2 for each of the three SoG groupings, it can be seen that the mean for |SoG| > 3° is lower than the mean for 1° < |SoG| < 3° which is lower or equal than the mean for |SoG| < 1°.

Table 2. Mean maple and ash sample strain-to-failure results sorted by SoG class.

SoG Class	Maple Strain-to-Failure	Ash Strain-to-Failure
<1°	0.026 ± 0.0043	0.027 ± 0.0025
1° < SoG < 3°	0.025 ± 0.0030	0.027 ± 0.0020
>3°	0.023 ± 0.0041	0.024

However, this interpretation of the data may be a stretch due to the limited number of samples in each of the SoG groupings that span the range of densities. Without more data, it is difficult to make a definitive conclusion by using only the current research data on the relationship between SoG and strain-to-failure. Thus, in the absence of more data, the available results suggest that density has a greater effect on the impact failure properties of the wood than SoG for the range of densities and SoGs considered in this study. However, once the wood has begun to fracture, SoG dominates the direction of crack propagation of the wood, thereby determining if the pieces breaking away from the bat are fairly blunt or spear-like in shape. This phenomenon is highlighted in Figure 8.

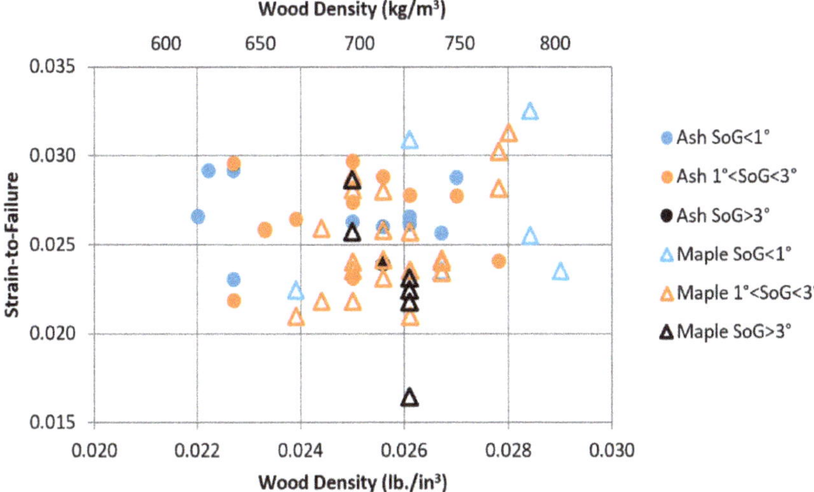

Figure 7. Strain-to-failure of ash samples impacted on the edge-grain and maple samples impacted on the face-grain. All samples are sorted by slope-of-grain (SoG).

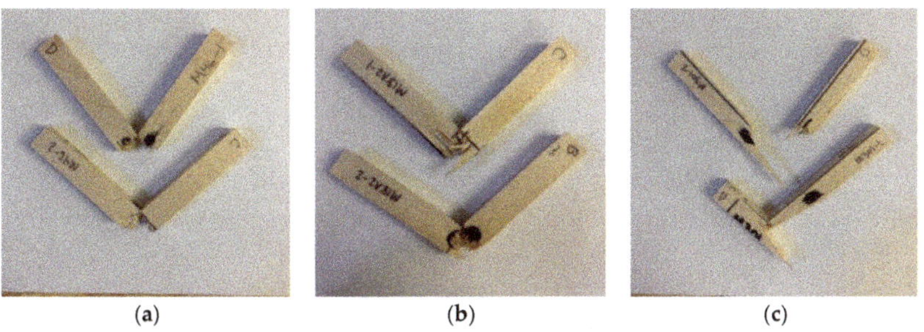

Figure 8. Charpy samples of varying SoG (**a**) |SoG| < 1°, (**b**) 1° < |SoG| < 3° and (**c**) |SoG| > 3°.

Figure 8 shows a maple wood species Charpy test sample from each of the three SoG groupings that were examined. The samples shown in Figure 8a have a density of 0.028 lb/in^3 (786 kg/m^3) and a SoG measuring 0.3°. It can be seen that the sample has not completely fractured through the width and this fracture surface is indicative of a brash wood failure. The samples in Figure 8b have a density of 0.026 lb/in^3 (709 kg/m^3) and a SoG measuring 1.7°. Theses samples exhibit more splintering of the wood at the fracture site in comparison to Figure 8a. The Figure 8b samples also did not completely split through the width. Figure 8c shows samples with a wood density of 0.026 lb/in^3 (722 kg/m^3) and a measured SoG of 3.7°. The high-SoG specimens split along the grain line during impact, thereby resulting in the formation of sharp edges on the fracture surface. This high-SoG failure propagation behavior is important to consider when discussing the issue of broken bats during gameplay because high-SoG wood is very likely to fail in a multi-piece mode.

3.3. Finite Element Model

The results of the quasi-static and dynamic test programs of ash and maple assisted in building a comprehensive set of the wood material properties that are needed for the LS-DYNA material model *MAT_WOOD (MAT_143). To investigate the credibility of a finite element model using this set of material properties, a finite element model of a Charpy Impact specimen was developed for analysis in LS-DYNA. The model was constructed of 10,000 solid elements and 12,221 nodes with a uniform mesh size of 0.0500 in. (0.127 cm). Previous modeling of the breaking of wood bats in LS-DYNA found that good correlation between models and experiments were best achieved when the *MAT_ADD_EROSION option for prescribing a strain-to-failure criterion was used in combination with the *MAT_WOOD (*MAT_143) material model [18]. Thus, the Charpy models were investigated using the *MAT_WOOD (*MAT_143) material model in combination with the *MAT_ADD_EROSION option. Once a high level of correlation was demonstrated for simulations of the Charpy test, the material properties and failure criterion were used to model baseball bat impacts and the outputs of these bat/ball impact simulations were compared to existing bat failure testing results.

3.4. Wood Material Cards

Material Type *MAT_WOOD (MAT_143) is the wood based material model available for use in LS-DYNA [19]. The user prescribes the material input parameters for moduli, strength and fracture properties of the wood species. For Card 2 in *MAT_WOOD (MAT_143), the Young's modulus was calculated using the linear equation for MOE versus wood density that was derived from the results of the FPL quasi-static testing on maple. To characterize the remaining moduli parameters available in *MAT_WOOD (MAT_143), the relationships published in table 5-1 of the Wood Handbook [7] were used as a guide. The strength input parameters for Card 3, for example, the tensile yield strength, were determined using the linear regression for MOR versus wood density that was derived from the results of the FPL quasi-static testing on maple to provide a value for the parallel tensile strength. These values are displayed in bold in Table 3. The remaining strength input parameters were derived based on the strength relationships published in table 5-3 in the Wood Handbook [7]. The input parameters for 0.0250 lb/in^3 (692.0 kg/m^3) maple wood are shown in Table 3. This approach was also used to determine the ash wood material properties as a function of density as listed in Table 3.

The Card 4 fracture input parameters for Mode I parallel fracture energy were based on the values published in tables 5–10 of the Wood Handbook [7] of 430 lbf/in^2-in$^{1/2}$ (480 KPa-m$^{1/2}$) as this failure mode is seen during a Charpy Impact test. The remaining parameters were left as determined by the FHWA study [11].

Table 3. MAT_WOOD (MAT_143) maple and ash card 2 and 3 inputs.

Material Property	LS-DYNA Material Variable	Units		Ash		Maple	
		US	SI	US	SI	US	SI
Parallel Normal Modulus	EL	lb/in^3	GPa	2,190,000	15.1	2,280,000	15.7
Perpendicular Normal Modulus	ET	lb/in^3	GPa	175,000.0	1.21	148,070.0	1.02
Parallel Shear Modulus	GLT	lb/in^3	GPa	238,438.0	1.64	252,858.0	1.74
Perpendicular Shear Modulus	GTR	lb/in^3	GPa	83,864.00	0.58	80,642.00	0.56
Poisson's Ratio	PR	—	—	0.440		0.476	
Parallel Tensile Strength	**XT**	**lb/in^3**	**MPa**	**24,474.00**	**169**	**22,513.00**	**155**
Parallel Compressive Strength	XC	lb/in^3	MPa	12,090.30	83.4	11,227.02	77.4
Perpendicular Tensile Strength	YT	lb/in^3	MPa	1942.400	13.4	2163.920	14.9
Perpendicular Compressive Strength	YC	lb/in^3	MPa	1892.600	13.1	2107.850	14.5
Parallel Shear Strength	SXY	lb/in^3	MPa	3116.300	21.5	3341.050	23.0
Perpendicular Shear Strength	SYZ	lb/in^3	MPa	4362.800	30.1	4677.470	32.3

3.5. Failure Criterion

An important aspect to the finite element modeling is demonstrating the ability of the model to capture the failure of the wood. To accomplish this failure in the model, a reliable failure criterion must be established and implemented into the modeling. The results of the dynamic material testing provided valuable insight into the strain-to-failure of the wood species and how the strain-to-failure does (or does not) vary as a function of wood density. Using these results, a relationship between wood density and failure strain was determined. This relationship for maple and ash wood is given in Equations (6) and (7), respectively. Incorporating the strain-to-failure into the modeling was done through the use of the *MAT_ADD_EROSION option which provides an element failure condition based on user-defined inputs. For the current study, the maximum principle strain at failure input was utilized.

$$Maple \ \varepsilon_f = 0.8986 \times \rho + 0.0016 \tag{6}$$

$$Ash \ \varepsilon_f = 0.1685 \times \rho + 0.0219 \tag{7}$$

3.6. Model Correlation

A finite element model of the Charpy Impact was analyzed in LS-DYNA and the correlation of the model with experimental results was examined. The model is shown in Figure 9 where the setup consists of a 5.00 × 0.50 × 0.50 in. (12.7 cm × 1.3 cm × 1.3 cm) (length × width × height) wood specimen impacted by a 10.0 ft-lb (13.6 N-m) Charpy hammer at 11.4 ft/s (3.47 m/s). The wood specimen rests on a pair of rigid support brackets. The element mesh is very fine, so explicitly showing the mesh would not be beneficial.

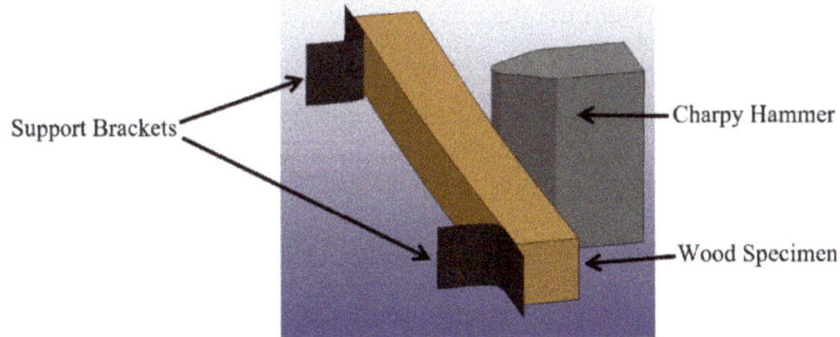

Figure 9. Charpy impact finite element model.

The simulation was processed using LS-DYNA R8.0 and postprocessed using LS-PrePost 4.1. To minimize the bias that may come with any anomalies associated with one specific test, the mean density of 0.026 lb/in^3 (719.7 kg/m^3) and the associated strain-to-failure of 0.025 (per Equation (6)) of the maple data set were used for this finite element model. The resulting maximum deflection of the model before failure occurs is 0.137 in. (0.348 cm), which is very close to the 0.138 in. (0.350 cm) maximum deflection of sample M307B, a 0.026 lb/in^3 (719.7 kg/m^3) density maple sample that failed at a strain of 0.0241. The model deflection of 0.137 in. (0.348 cm) falls within the mean deflection plus one standard deviation of 0.12–0.159 in. (0.308–0.404 cm) of the entire maple sample set. The model can be seen in Figure 10 with a high-speed video image for comparison.

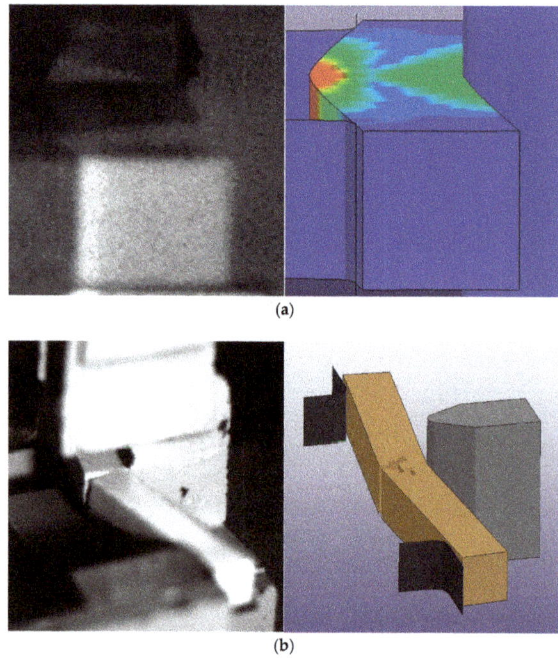

Figure 10. Charpy Impact of maple sample: (**a**) High-speed video image and (**b**) Finite element model showing maximum strain contours (**top**) and crack propagation (**bottom**).

A series of finite element simulations of the Charpy impact tests was completed for maple and ash specimens and a span of wood densities for each of the two wood species. Summaries of the comparison of maximum deflection, that is, deflection just before fracture, of these maple and ash Charpy impact models with the corresponding test sample IDs are presented in Table 4, respectively. There is excellent correlation between the maximum deflection values between the experimental data and the model results.

Table 4. Charpy Impact model maple and ash wood deflection correlation results.

Sample ID	Density lb/in^3 (kg/m^3)	Maximum Deflection in. (cm)		% Difference
		Experiment	FE Model	
M35B10	0.024 (664.3)	0.125 (0.318)	0.126 (0.320)	0.80%
M36C	0.025 (692.0)	0.133 (0.338)	0.132 (0.335)	0.75%
M307	0.026 (719.7)	0.138 (0.350)	0.137 (0.348)	0.72%
M306	0.027 (747.4)	0.152 (0.386)	0.149 (0.378)	1.97%
A44B10	0.023 (636.6)	0.136 (0.345)	0.138 (0.351)	1.47%
A79A10	0.025 (692.0)	0.148 (0.376)	0.147 (0.373)	0.68%
A41A10	0.026 (719.7)	0.148 (0.376)	0.148 (0.376)	0.00%

3.7. Bat Model

The methods described for the modeling of the maple wood in the Charpy impact simulations were applied to the finite element modeling of bat/ball impacts. A finite element model of a popular professional bat profile that was previously tested in the Bat Durability Test System (Automated Design Corporation) at the UMLBRC was built for subsequent analysis in LS-DYNA. The finite element model of the bat was constructed containing 130,554 solid elements and 137,548 nodes with a mesh size of approximately 0.2 in. (0.4920 cm). This maple bat model was 34 in. (86.4 cm) in length with a prescribed wood density of 0.0250 lb/in^3 (692.0 kg/m^3) to represent a 31 oz. (0.879-kg) bat, thereby matching the profile, length and weight of the lab tested maple bat. The finite element model of the baseball that was used in this research was constructed in LS-PrePost. The geometry of the ball model is a sphere of 1.4 in. (3.6 cm) radius and consists of 12,096 solid elements and 12,589 nodes with a uniform mesh size of approximately 0.2 in. (0.556 cm). The LS-DYNA *MAT_VISCOELASTC (*MAT_006) was used to define the material behavior of the ball. The baseball material properties are summarized in Table 5 [20].

Table 5. Ball material properties.

Density (blobs/in^3) [Kg/m^3]	Bulk Modulus (lb/in^2) [N/m^2]	Initial Shear Modulus (lb/in^2) [N/m^2]	Long-Term Shear Modulus (lb/in^2) [N/m^2]	β
7.230 × 10^{-5} [772.5]	13,500 [93.1]	5000 [34.5]	1080 [7.45]	20,000

The finite element model of the Bat Durability Test System consisted of the rotating back plate and the rollers that grip the bat. The LS-DYNA *MAT_ELASTIC (*MAT_001) was used to implement the properties for each of the rollers. The LS-DYNA *MAT_RIGID (*MAT_020) with aluminum material properties for the elastic modulus and Poisson's ratio was used for the plate. The material properties of the rollers and "rigid" aluminum plate are summarized in Table 6.

Table 6. ADC roller and aluminum plate material properties.

Part	LS-DYNA Material Model	Density (blobs/in^3) [Kg/m^3]	E (lb/in^2) [N/m^2]	Poisson's Ratio
Top Rollers	#1 Isotropic Elastic	2.5880 × 10^{-5} [276.5]	750,000 [5.171 × 10^9]	0.3
Bottom Rollers	#1 Isotropic Elastic	2.5880 × 10^{-5} [276.5]	75,000 [5.171 × 10^8]	0.3
Plate	#20 Rigid	2.5907 × 10^{-4} [2768]	10,000,000 [68.95 × 10^9]	0.3

An impact velocity of 143 mph (230 kph) at a location 14.0 in. (35.6 cm) from the barrel tip was prescribed in the model to mimic the lab test conditions. Surface-to-surface contact was defined between the baseball and the bat models. The strain-to-failure used in the model was 0.0241 and was derived from the relationship between maple wood density and strain-to-failure based on the results of the Charpy impact testing (Equation (6)). A comparison of a high-speed image taken of the bat profile during durability testing and the finite element model are presented in Figure 11. The model shows excellent correlation for the mode of fracture (single-piece failure), location of fracture initiation (~15 in. [38 cm] from the knob) and the extent of the crack propagation along the length of the bat (~6 in. [15 cm] in total length).

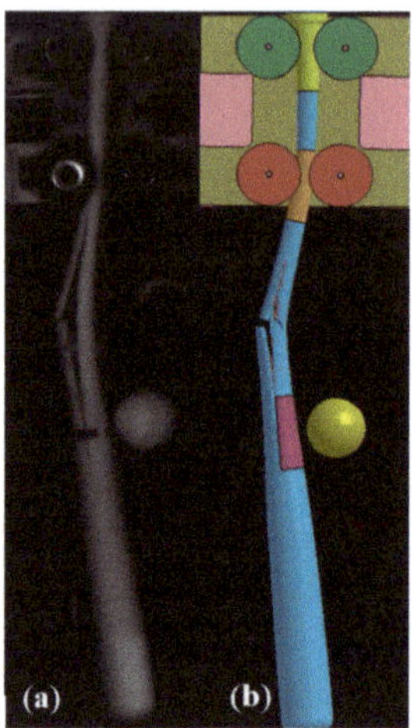

Figure 11. 34-in. (86.4 cm) long, 31-oz. (0.879 kg) professional maple bat profile impacted at 143 mph (230 kph): (**a**) Lab test and (**b**) Finite element model.

In an effort to further explore the capabilities of this maple bat model, the impact velocity was varied to observe the bat response at different speeds. Through ramp-up durability testing at the UMLBRC, it is known that bat profiles will be able to withstand impacts up to a distinct failure threshold and then fracture in either a single-piece or multi-piece failure mode will occur.

Three impact velocities were chosen, that is, 120, 130 and 145 mph (193, 209 and 233 kph) for examining how the response of the bat varies with impact speed. The wood properties and failure strain were unchanged from the 34 in. (86.4 cm) 31 oz. (0.879 kg) bat as shown in Figure 11. The post-impact results of the modeling are shown in Figure 12. It can be seen in this figure that there is a transition from single-piece failure to multi-piece failure somewhere between 130 and 145 mph (209 and 233 kph). By using 5 mph (8 kph) incremental changes in the impact velocity, a good estimate of the impact speed associated with the single-piece failure and multi-piece failure thresholds can be determined.

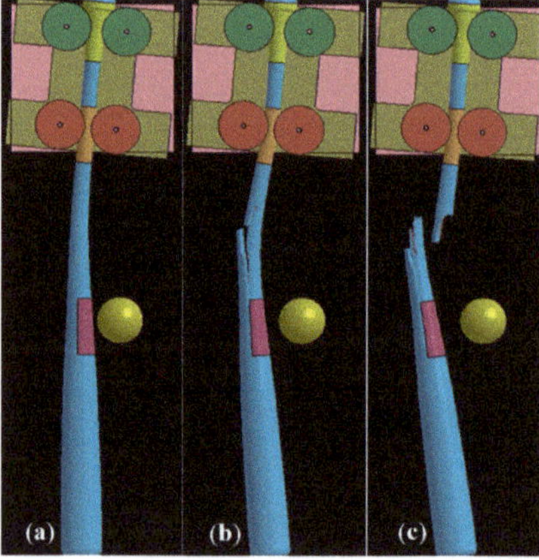

Figure 12. Bat model failure modes at varying impact velocities; (**a**) No failure at 120 mph (193 kph), (**b**) Single-piece failure at 130 mph (209 kph) and (**c**) Multi-piece failure at 145 mph (233 kph).

To explore the respective roles that impact density and strain-to-failure play in the subsequent failure mode of the bat, the density was varied from 0.0245–0.0265 lb/in^3 (678–734 kg/m^3) in

0.00025 lb/in^3 (6.9 kg/m^3) increments with the material properties of the bat prescribed based on the relationships defined in Equations (1) and (2). At each increment the strain-to-failure defined in the *MAT_ADD_EROSION card of the model was scaled according to the relationship defined in Equation (6). The bat was impacted at a velocity of 145 mph (233 kph) and the subsequent bat response was analyzed. Figure 13 shows the failure mode of a C243 profile at the three different density levels. The density of the wood plays a significant role in the subsequent failure mode of the bat, as wood material properties such as stiffness and strength increase with increasing density. As the density of the bat is increased, the failure of the bat transitions from multi-piece failure to single-piece failure and finally reaches a level where no failure occurs for a velocity of 145 mph (233 kph) and the bat survives the impact. This result shows that bats comprised of higher density wood are more durable and less susceptible to failure than those with lower densities.

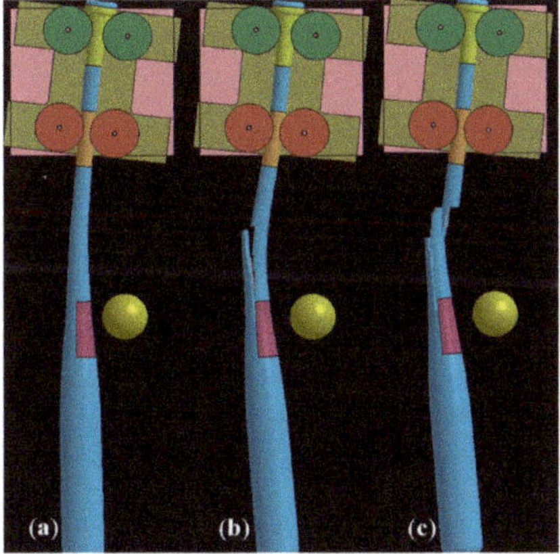

Figure 13. C243 profile impacted at 145 mph (233 kph) at density levels of (**a**) 0.0265 lb/in^3 (733 kg/m^3), (**b**) 0.0255 lb/in^3 (706 kg/m^3) and (**c**) 0.0245 lb/in^3 (678 kg/m^3).

4. Conclusions

Charpy impact testing at strain rates comparable to what a wood bat experiences during a bat/ball collision was completed to characterize the failure energy and strain-to-failure as a function of density and SoG for ash and maple. The failure energy increased with increasing wood density for the maple wood species. The failure energy exhibited a slight decrease with increasing wood density for the ash samples. The strain-to-failure increased with increasing density for maple but the strain-to-failure remained constant over the range of densities considered in this study for ash. The relatively flat slope of the ash data suggests that ash-bat durability is less sensitive to wood density than maple bats. Ash exhibited a higher strain-to-failure in edge-grain loading in comparison to face-grain loading and maple exhibited a higher strain-to-failure in face-grain loading in comparison to edge-grain loading. These results support the WBBS requirement that ash bats be impacted on the edge grain and maple bats be impacted on the face grain. Maple exhibited a lower failure energy, a lower max-deflection and a lower strain-to-failure for both edge and flat-grain loadings relative to ash. Maple exhibited a higher failure energy when loaded in the face-grain direction compared to edge-grain loading. The available SoG results suggest that density has a greater effect on the impact failure properties of the wood than SoG. However, once the wood has begun to fracture, SoG is the primary influence of the direction of crack propagation of the wood, thereby determining if the pieces breaking away from the bat are fairly blunt or spear-like in shape.

The characterization data were subsequently used in finite element models of a Charpy test and of a bat/ball collision. The series of finite element models of the Charpy impact tests spanned the range of ash and maple densities used in the test program. The finite element models showed excellent correlation with the experimental data for the same combinations of wood species and density. A finite element model of a maple bat/ball impact showed excellent correlation with a lab test of the same conditions.

Author Contributions: Conceptualization, J.S., P.D. and D.K.; Data curation, J.F.-S., J.S. and P.D.; Formal analysis, J.F.-S.; Funding acquisition, J.S.; Investigation, J.F.-S.; Methodology, J.F.-S. and J.S.; Project administration, P.D.; Resources, P.D.; Software, J.F.-S. and J.S.; Supervision, J.S. and P.D.; Validation, J.F.-S., J.S., P.D. and D.K.; Visualization, J.F.-S.; Writing—original draft, J.F.-S.; Writing—review & editing, J.S., P.D. and D.K.

Funding: This research was funded in part by the Office of the Commissioner of Baseball and the MLB Players Association.

Acknowledgments: This material is based upon the work supported by the Office of the Commissioner of Baseball and the MLB Players Association. Any opinions, findings and conclusions or recommendations expressed in this material are those of the authors and do not necessarily reflect the views of the Office of the Commissioner of Baseball and the MLB Players Association.

Conflicts of Interest: The authors declare no conflict of interest. The funding sponsors had no role in the design of the study; in the collection, analyses, or interpretation of data; in the writing of the manuscript and in the decision to publish the results.

References

1. Ruggiero, E. Investigating Baseball Bat Durability Using Experimental and Finite Element Modeling Methods. Master's Thesis, University of Massachusetts Lowell, Lowell, MA, USA, 2013.
2. Nicholls, R.L.; Miller, K.; Elliott, B.C. Numerical analysis of maximal bat performance in baseball. *J. Biomech.* **2006**, *39*, 1001–1009. [CrossRef] [PubMed]
3. Mustone, T.J.; Sherwood, J.A. Using LS-DYNA to Characterize the Performance of Baseball Bats. In Proceedings of the LS-Dyna Users Conference, Detroit, MI, USA, 21–22 September 1998.
4. Shenoy, M.M.; Smith, L.V.; Axtell, J.T. Performance assessment of wood, metal and composite baseball bats. *Compos. Struct.* **2001**, *52*, 397–404. [CrossRef]
5. Smith, L.V.; Hermanson, J.C.; Rangaraj, S.; Bender, D.A. Dynamic finite element analysis of wood baseball bats. In Proceedings of the 1999 ASME Bioengineering Conference, Big Sky, MT, USA, 16–20 June 1999; pp. 629–630.
6. Fortin-Smith, J.; Ruggiero, E.; Drane, P.; Sherwood, J.; Kretschmann, D. A Finite Element Investigation of the Relationship between Bat Taper Geometry and Bat Durability. *J. Sports Eng. Technol.* **2017**, in press. [CrossRef]
7. Kretschmann, D.E. *Wood handbook—Wood as an Engineering Material*; General Technical Report FPL-GTR-190; U.S. Department of Agriculture, Forest Service, Forest Products Laboratory: Madison, WI, USA, 2010.
8. Kretschmann, D.E.; Bridwell, J.J.; Nelson, T.C. Effect of changing slope of grain on ash, maple, and yellow birch in bending strength. In Proceedings of the WCTE 2010, World Conference on Timber Engineering, Riva del Garda, Trento, Italy, 20–24 June 2010.
9. Bertrand, D.; Bourrier, F.; Olmedo, I.; Brun, M.; Berger, F.; Limam, A. Experimental and numerical dynamic analysis of a live tree stem impacted by a Charpy pendulum. *Int. J. Solids Struct.* **2013**, *50*, 1689–1698. [CrossRef]
10. Reid, S.R.; Peng, C. Dynamic uniaxial crushing of wood. *Int. J. Impact Eng.* **1997**, *19*, 531–570. [CrossRef]
11. Murray, D.; Reid, J.D.; Faller, R.K.; Bielenberg, B.W.; Paulsen, T.J. *Evaluation of LS-DYNA Wood Material Model 143*; Publication No. FHWA-HRT-04-096; Federal Highway Administration: Washington, DC, USA, 2005.
12. U.S. Department of Transportation Federal Highway Administration. *Manual for LS-DYNA Wood Material Model 143 2007*; U.S. Department of Transportation Federal Highway Administration: Washington, DC, USA, 2007.
13. ASTM International. *ASTM International 2010 ASTM D6110-10. Standard Test Method for Determining the Charpy Impact Resistance of Notched Specimens of Plastics*; ASTM International: West Conshohocken, PA, USA, 2010.
14. Fortin-Smith, J.; Ruggiero, E.; Drane, P.; Sherwood, J.; Kretschmann, D. Wood Baseball Bat Slope of Grain Durability Finite Element Modeling. *Appl. Sci.* **2018**. submitted for review.
15. Ruggiero, E.; Fortin-Smith, J.; Sherwood, J.; Drane, P.; Kretschmann, D. Development of a Protocol for Certification of New Wood Species for Making Baseball Bats. *J. Sports Eng. Technol.* **2017**, in press.

16. Carrasco, E.V.M.; Mantilla, J.N.R. Applying Failure Criteria to Shear Strength Evaluation of Bonded Joints According to Grain Slope under Compressive Load. *Int. J. Eng. Technol.* **2013**, *13*, 19–25.
17. Hankinson, R.L. Investigation of crushing strength of spruce at varying angles of grain. *Air Serv. Inf. Circ.* **1921**, *3*, 130.
18. Ruggiero, E.; Sherwood, J.; Drane, P.; Duffy, M.; Kretschmann, D. Finite Element Modeling of Wood Bat Profiles for Durability. *Proc. Eng.* **2014**, *34*, 427–432. [CrossRef]
19. Livermore Software Technology Corporation. *LS-DYNA KEYWORD USER'S MANUAL VOLUME II—Material Models*; Livermore Software Technology Corporation: Livermore, CA, USA, 2016.
20. Connelly, T.; University of Massachusetts Lowell, Lowell, MA, USA. Personal Communication, 2011.

Article

A Finite Element Investigation into the Effect of Slope of Grain on Wood Baseball Bat Durability

Joshua Fortin-Smith [1], James Sherwood [1,*], Patrick Drane [1], Eric Ruggiero [1], Blake Campshure [1] and David Kretschmann [2,†]

[1] Baseball Research Center, Mechanical Engineering Department, University of Massachusetts Lowell, 1 University Avenue, Lowell, MA 01854, USA
[2] U.S. Forest Products Laboratory, US Forest Service, 1 Gifford Pinchot Drive, Madison, WI 53726, USA
[*] Correspondence: James_Sherwood@uml.edu; Tel.: +1-978-934-3313
[†] Retired.

Received: 30 July 2019; Accepted: 30 August 2019; Published: 7 September 2019

Abstract: Bat durability is defined as the relative bat/ball speed that results in bat breakage, i.e., the higher the speed required to initiate bat cracking, the better the durability. In 2008, Major League Baseball added a regulation to the Wooden Baseball Bat Standards concerning Slope-of-Grain (SoG), defined to be the angle of the grain of the wood in the bat with respect to a line parallel to the longitudinal axis of the bat, as part of an overall strategy to reverse what was perceived to be an increasing rate of wood bats breaking into multiple pieces during games. The combination of a set of regulations concerning wood density, prescribed hitting surface, and SoG led to a 30% reduction in the rate of multi-piece failures. In an effort to develop a fundamental understanding of how changes in the SoG impact the resulting bat durability, a popular professional bat profile was examined using the finite element method in a parametric study to quantify the relationship between SoG and bat durability. The parametric study was completed for a span of combinations of wood SoGs, wood species (ash, maple, and yellow birch), inside-pitch and outside-pitch impact locations, and bat/ball impact speeds ranging from 90 to 180 mph (145 to 290 kph). The *MAT_WOOD (MAT_143) material model in LS-DYNA was used for implementing the wood material behavior in the finite element models. A strain-to-failure criterion was also used in the *MAT_ADD_EROSION option to capture the initiation point and subsequent crack propagation as the wood breaks. Differences among the durability responses of the three wood species are presented and discussed. Maple is concluded to be the most likely of the three wood species to result in a Multi-Piece Failure. The finite element models show that while a 0°-SoG bat is not necessarily the most durable configuration, it is the most versatile with respect to bat durability. This study is the first comprehensive numerical investigation as to the relationship between SoG and bat durability. Before this numerical study, only limited empirical data from bats broken during games were available to imply a qualitative relationship between SoG and bat durability. This novel study can serve as the basis for developing future parametric studies using finite element modeling to explore a large set of bat profiles and thereby to develop a deeper fundamental understanding of the relationship among bat profile, wood species, wood SoG, wood density, and on-field durability.

Keywords: baseball; bat; durability; finite element; impact; slope of grain; wood

---

1. Introduction

In the early years of baseball, bats were made from a wide variety of wood species as players explored various woods in their quest to find the perfect wood. Eventually, northern white ash became the wood of choice as it had the best combination of density, stiffness, and impact resistance properties. For example, hickory has better impact resistance than the northern white ash, but it is denser than ash, thereby making for a relatively heavy bat. In the late 19th century, the Hillerich family in Louisville, Kentucky cemented the central role of ash bats in baseball when it began manufacturing northern white ash bats for a number of professional teams [1].

Ash remained the wood of choice until the late 1990s when maple bats were introduced to players by Sam Bat, and a small number of players transitioned from ash to maple [2]. As the popularity of maple bats increased, the rate at which bats were breaking into multiple pieces during games was perceived to be increasing as well. However, there were no data to support whether this phenomenon was real or perceived. To obtain concrete data on the phenomenon, Major League Baseball (MLB) in cooperation with the Major League Baseball Players Association (MLBPA) authorized the collection of all bats that broke in games during a $2\frac{1}{2}$-month portion of the 2008 MLB regular-season. Visual inspection of these broken bats concluded that the slope-of-grain (SoG) of the wood in each of the bats played a major role in the likelihood of a bat breaking into multiple fragments [3]. Slope-of-grain is defined to be the angle of the grain of the wood in the bat with respect to a line parallel to the longitudinal axis of the bat.

As a result of the SoG observations from the field-collection bats and the understanding of the mechanical behavior of maple from a wood-science perspective, the wood-science experts on the team recommended new regulations be implemented after the 2008 season with respect to (1) setting the allowable range of SoG to be $\pm 3°$ and (2) changing the hitting surface for maple bats from edge grain to face grain. Before limiting the allowable range of SoG to be $\pm 3°$, many bat manufacturers were not closely monitoring SoG, and therefore, allowing wood with SoGs of up to $\pm 10°$ to be used in bats.

After the 2009 season, additional regulations were implemented to restrict the minimum allowable wood density to 0.0245 lb/in^3 (678 Kg/m^3) for maple wood species that could be used to manufacture baseball bats. Before this limitation was implemented, any species of maple was allowed to be used to make a wood bat. Without a list of specific maple species, maple species with poor impact properties were sometimes being used. After the introduction of the strict limitation to a single species of maple and greater awareness and compliance with the new regulations, the MPF rate dropped an additional 15–20% between the 2009 and 2010 seasons [4]. Each of the new regulations had the benefit that the improvements in bat durability came with changes that were fairly transparent to the players, i.e., there were no bat geometries or popular wood species that were prohibited from use.

The combination of (1) setting the allowable range for SoG to be $\pm 3°$, (2) changing the impact surface from the edge to face grain for maple, and (3) the restriction on the minimum allowable wood density of 0.0245 lb/in^3 (678 kg/m^3) for the wood species of maple resulted in a measurable decrease of ~65% in the MPF rate per game [5]. Of these three changes, the only one that has an option to offer any additional drop in the MPF rate without any increase in bat weight is a further tightening of the allowable SoG from $\pm 3°$. To explore if and how a further tightening of the allowable SoG could improve upon the MPF rate, the durability of sets of bats sorted by SoG, e.g., sorted into four SoG groupings, specifically (1) $|SoG| \leq 1°$, (2) $1° < |SoG| \leq 2°$, (3) $2° < |SoG| \leq 3°$, and (4) $|SoG| > 3°$, could be tested in the lab. However, such an approach would be difficult due to the inherent variability of wood and the lack of control over the factors unrelated to SoG, such as density or the ultimate strength of the wood, thereby making it challenging to draw conclusive trends. In addition, it would also be expensive with respect to the large number of bats that would need to be explored for such a study, as well as the time required to characterize the durability of all bats in the study and subsequent examination of all bat failures. An alternative and cost-effective approach is to use the finite element method to investigate how the failure mode changes as a function of SoG for a range of representative on-field relative bat/ball speeds and impact locations [6–8].

In the current paper, a popular bat profile is used in a parametric study to investigate the relationship between SoG and bat durability. During a game, the relative bat/ball impact velocity can range from 90–180 mph (145–290 kph) and is a consequence of the combination of the pitch speed, swing speed, and impact location along the length of the bat. For the purposes of this paper, bat durability is defined as the relative bat/ball speed that results in bat breakage, i.e., the higher the speed required to initiate bat cracking, the better the durability. The finite element models of this bat profile are analyzed using LS-DYNA where the variables in the parametric study include (1) SoG—ranging between ±10°, (2) Wood species—Ash, Maple, and Yellow Birch, (3) two impact locations—2.0 in. (5.1 cm) and 14.0 in. (35.6 cm) as measured from the tip of the barrel, and (4) relative bat/ball impact speeds between 90 and 180 mph (145–290 kph). This paper presents the first comprehensive numerical investigation into the relationship between SoG and bat durability. Prior to this numerical study, limited empirical data from bats broken during games and in-laboratory studies where SoG was not a controlled variable were available to imply a qualitative relationship between SoG and bat durability. The results of this study can serve as the basis for developing future parametric studies using finite element modeling to explore a large set of bat profiles and thereby to develop a deep fundamental understanding of the relationship among bat profile, wood species, wood SoG, wood density, and on-field durability.

2. Background

Baseball bats are comprised of four major regions, as shown in Figure 1. The barrel is the intended ball impact region. The barrel of a typical 34.0 in. (86.4 cm) long baseball bat extends from the tip of the barrel to 10.0–12.0 in. (25.4–30.5 cm) down the length of the bat. The taper region is where the diameter of the bat transitions from the large diameter of the barrel to the small diameter of the handle. The taper region of a 34.0-in. (86.4 cm) bat runs from the end of the barrel section to about 20.0 in. (50.8 cm) from the tip of the barrel. Manufacturers are required to place their company label on this region of the bat, and the location of the label should be located 90° away from the intended hitting surface of the bat such that a batter can follow the protocol of hitting with the label up or label down. The handle of a baseball bat is the section that a player grips when swinging. The handle of a 34.0 in. (86.4 cm) bat extends from the end of the taper region to about 33 in. (84 cm) from the tip of the barrel. The final ~1 in. (~2.5 cm) of a wood baseball bat is the knob, which has a diameter a bit larger than that of the grip area so as to help the player maintain his hold on the bat during the swing.

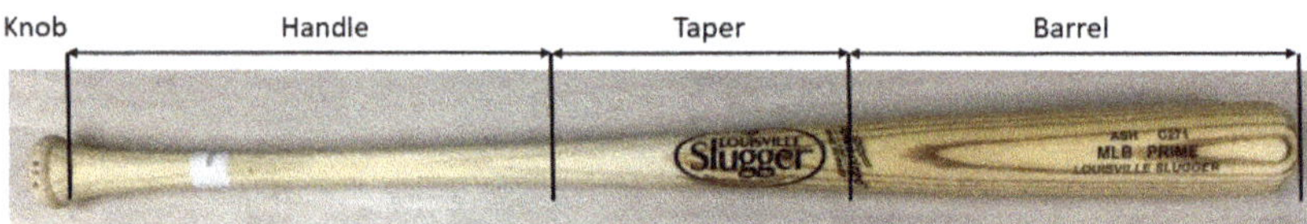

Figure 1. Ash baseball bat with logo on the face-grain surface of the bat. The knob, handle, taper and barrel sections of the bat are identified.

Upon the wide acceptance of northern white ash as the wood of choice, the players also settled on holding the bat such that the ball would impact the edge grain of the wood as opposed to the face grain (Figure 2). As a result, bat manufacturers located their logo on the face grain of the bat, i.e., 90° away from the edge grain. Thus, players would commonly hit with the logo up or logo down. Figure 1 shows an example of the location of the Louisville Slugger logo on the face-grain surface of an ash bat.

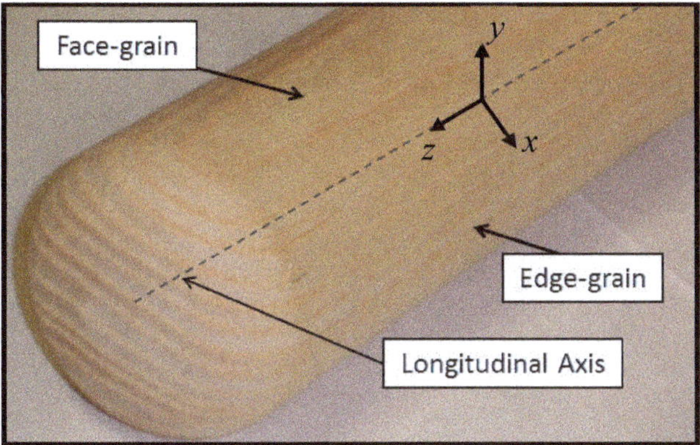

Figure 2. Face-grain and edge-grain surfaces of an ash wood baseball bat.

Figure 3 shows the respective directions of positive and negative SoGs for a label-up orientation of the bat and the ball traveling from the pitcher to home plate. In Figure 3, the bat is shown situated in a finite element model of the bat durability testing system at the University of Massachusetts Lowell Baseball Research Center (UMLBRC).

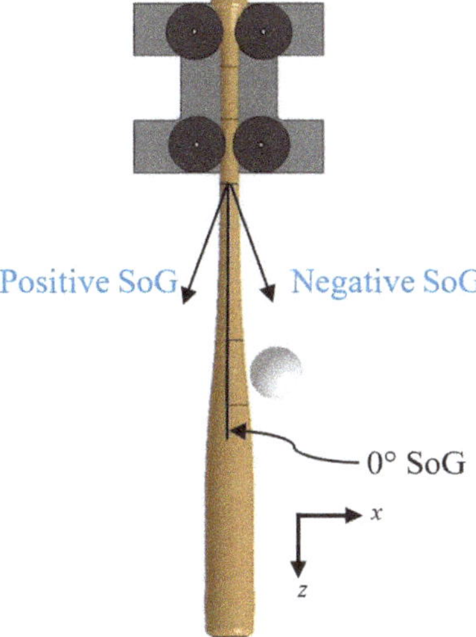

Figure 3. Finite element model of a baseball and the bat in the test fixture. The annotations denote the directions of positive, negative and 0° SoGs.

An SoG of 0°, i.e., the grain of the wood is aligned with the long axis of the bat, is alleged to be the optimal condition for bat durability. As the SoG deviates from 0°, the effective break strength, i.e., the modulus of rupture, of the wood decreases [6,9–11]. Hankinson's equation can be used to quantify the effect of SoG on wood mechanical properties,

$$N = \frac{PQ}{P\sin^n \theta + Q\cos^n \theta},$$ (1)

where N is the strength at an angle θ to the fiber direction, P is the strength perpendicular to the fiber, Q is the strength parallel to the fiber, and n is an empirically determined constant [10]. Different properties follow different curves (Figure 4), and these curves are average relationships developed across many species. The Wood Handbook states that the modulus of rupture (MOR) falls very close to the curve Q/P = 0.1 and n = 1.5. Using these values, the Hankinson Equation for a 3° SoG results in $N = 0.895$, meaning that the strength of a 3° SoG specimen is 89.5% of a specimen with a 0° SoG.

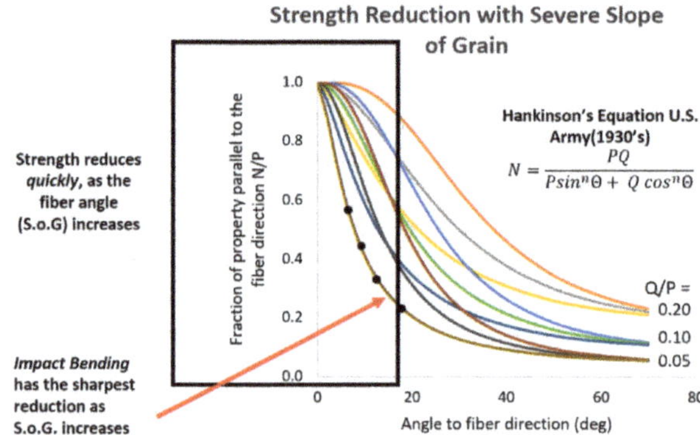

Figure 4. Illustration of Hankinson's equation [10].

The MOR curve in Figure 4 shows that an SoG of ±10° will result in a 60% drop in MOR—thus, the motivation to tighten the allowable SoG range to be ±3°. The motivation for the change in hitting surface was based on the knowledge that the impact resistance of maple, a diffuse porous wood, is better on the face grain than on the edge grain. This behavior is opposite to the impact durability of northern white ash, a ring porous wood species, which demonstrates superior durability when impacted on the edge grain face [12]. Recent Charpy testing of wood corroborated these differences in the hitting surfaces for ash and maple [6]. The tighter tolerance on the allowable SoG and the change in the hitting surface for maple from the edge to face grain for the 2009 season contributed to the rate of MPFs decreasing by 30% from what was observed in 2008 [13].

3. Materials and Methods

In this section, the basic topics which are relevant to the current finite element modeling studies are presented. These topics include a discussion on the lab testing for wood-bat durability, the single-piece failure (SPF) and a multi-piece failure (MPF) modes, and basic material properties for the ash, maple and yellow birch wood species.

3.1. Wood Bat Durability Testing

The UMLBRC uses a Bat Durability System (Automated Design Corporation, Romeoville, IL, USA) (Figure 5a,b) to investigate bat durability. Two pairs of compliant rollers hold a stationary bat and simulate the grip of the batter. At the knob end, the top set of rollers are firm to hold the bat in place while the bottom roller set is soft rubber and is loosely touching the bat handle. Previous work concluded that this setup in the Bat Durability System was representative of a player's grip of the baseball bat in the field [14]. A burp air cannon shoots baseballs toward the bat at speeds up to 200 mph (322 kph). The position of the bat can be raised or lowered such that the bat/ball impact can be programmed to occur anywhere along the length of the bat. A typical bat durability test commences with a speed known to be below the bat breaking speed for the axial position of interest, e.g., a hit on the taper that is 14.0 in. (35.6 cm) from the tip of the barrel (inside-pitch condition). The speed is then increased in increments of ~5 mph (~8 kph) on each subsequent hit. The test ends after the bat breaks, and the speed that induced either the MPF or SPF is used to quantify the bat durability. For this investigation, fatigue effects do not need to be considered because the bat breaking occurs within a range of 1 to 15 impacts. A detailed description of the bat durability testing process is given in Ruggiero et al. [15,16].

3.2. Modes of Wood Bat Failure

Bat breaks are classified as either a single-piece failure (SPF) or multi-piece failure (MPF). An SPF of a wood baseball bat occurs when the bat cracks but does not break into multiple large pieces. A large piece is defined to weigh 1.00 oz. (28.4 g) or more. An SPF type of failure is preferred over an MPF because the bat remains intact, and no fragments of the bat leave the batter's hands. Figure 5c shows an example of an SPF, for which a crack initiated but did not propagate either fully across the diameter of the bat or fully down the length of the bat. An MPF of a wood baseball bat occurs when the bat breaks into two or more large pieces. This type of failure is undesirable, as large and potentially sharp fragments can split from the bat and fly into the field or into spectator areas. Figure 5d shows an example of an MPF where a crack propagated down the length of the bat and almost entirely across the width of the bat along a diagonal line.

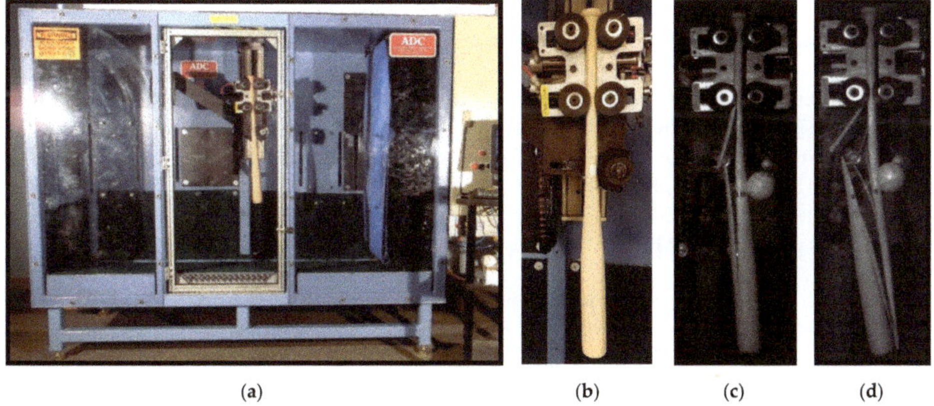

(a)   (b)   (c)   (d)

Figure 5. The testing system is shown with a wood bat hanging in machine and ready for impact: (**a**) the Bat Durability System, (**b**) Close-up view of bat in Bat Durability System, and examples of (**c**) SPF and (**d**) MPF.

3.3. Finite Element Models

All finite element analyses were completed using LS-DYNA (Livermore Software Technology Corporation, Livermore, CA, USA), and all finite element models used for this research were constructed using HyperMesh (Altair Engineering, Inc. Troy, MI, USA) and post-processed in LS-Prepost (Livermore Software Technology Corporation. Livermore, CA, USA). The finite element models of (1) a popular bat profile used by professional players (C243), (2) a baseball, and (3) the bat fixture of the Bat Durability System were constructed individually and then combined into a full model of a bat being tested at the UMLBRC for durability (Figure 6) [6]. Models were analyzed for a range of impact velocities that spanned from 90 mph to 180 mph (145 kph to 290 kph) in 5 mph (8 kph) increments, three wood species (ash, maple, and yellow birch) and wood SoGs between ±10° to investigate and to quantify how the SoG influences bat durability for variations in wood species and bat/ball impact speeds. The SoG's used in the models were −10°, −5°, −3°, −2°, −1°, 0°, 1°, 2°, 3°, 5°, and 10°. To validate the finite element models, a series of mechanical tests were performed. Four-point bending tests of the maple and ash wood were done to obtain elastic moduli and strength properties. Charpy testing was conducted to determine the strain-to-failure as a function of wood density [6,7].

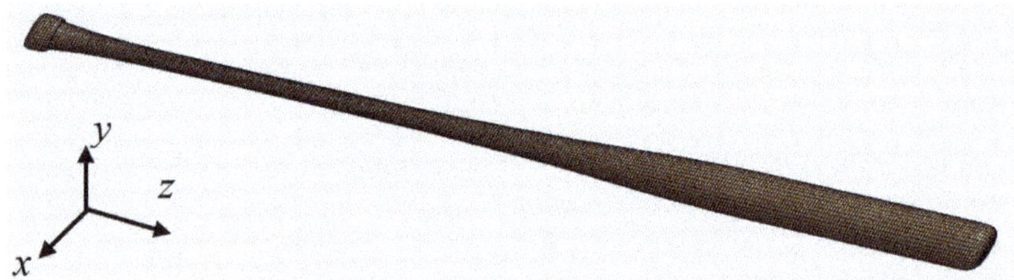

Figure 6. Finite element model of a C243 bat.

The bat model was constructed to the geometry of an uncupped C243 baseball bat profile. The finite element model was constructed using 8-noded brick elements with single Gauss-point integration. The LS-DYNA *MAT_WOOD (material model #143) definition has previously been demonstrated to be an effective material model for the simulation of wood structures under dynamic loading [17] and is used in this study for modeling the material behaviors of ash, maple, and yellow birch. The geometry of the C243 finite element model is shown in Figure 6. This model of a C243 bat profile is comprised of 179533 nodes and 159984 elements, with a mean mesh size of 0.2 in (0.4920 cm). This element size was concluded by completing a convergence study.

For this study, the wood density and specific wood-species material properties of the bat models were held constant throughout all of the models. The material properties for each wood species as they are provided in the LS-DYNA Keyword input file came from empirical testing and are summarized in Table 1. All models were prescribed a density of 0.0245 lb/in^3 (678 Kg/m^3), thereby resulting in a 33.0-oz (936-g) bat. The 0.0245 lb/in^3 (678 Kg/m^3) density was chosen because it is currently the lowest density that is permitted by MLB for maple bats.

Table 1. Summary of wood species material properties.

Material Property	LS-DYNA Material Variable	Ash	Maple	Yellow Birch
Parallel Normal Modulus	EL lb/in^3 (GPa)	2,130,000 (14.7)	2,260,000 (15.6)	2,340,000 (16.1)
Perpendicular Normal Modulus	ET lb/in^3 (GPa)	170,100.0 (1.17)	146,771.0 (1.01)	117,000.0 (0.81)
Parallel Shear Modulus	GLT lb/in^3 (GPa)	231,761.0 (1.60)	250,640.0 (1.73)	159,120.0 (1.10)
Perpendicular Shear Modulus	GTR lb/in^3 (GPa)	81,515.00 (0.56)	75,691.00 (0.52)	53,493.55 (0.37)
Poisson's Ratio	PR	0.440	0.476	0.451
Parallel Tensile Strength	XT lb/in^3 (MPa)	23,782.00 (164)	22,011.00 (152)	23,677.00 (163)
Parallel Compressive Strength	XC lb/in^3 (MPa)	11,748.00 (81.0)	10,977.00 (75.7)	11,638.30 (80.2)
Perpendicular Tensile Strength	YT lb/in^3 (MPa)	1888.000 (13.0)	2115.700 (14.6)	1310.600 (9.04)
Perpendicular Compressive Strength	YC lb/in^3 (MPa)	1839.100 (12.7)	2060.900 (14.2)	1381.800 (9.53)
Parallel Shear Strength	SXY lb/in^3 (MPa)	3028.200 (20.9)	3266.600 (22.5)	2678.100 (18.5)
Perpendicular Shear Strength	SYZ lb/in^3 (MPa)	4239.500 (29.2)	4573.300 (31.5)	3749.300 (25.9)

The SoG of the wood was varied between −10° and +10°. One of the required inputs for the *MAT_WOOD material model in LS-DYNA is the specification of the direction cosines (AOPT values) for the orientation of the wood [18]. These direction cosines allow for easily prescribing the SoG of the wood and have been shown to reliably predict impact velocity of bat failure, location of bat failure, and a mode of bat failure [18]. Both positive and negative wood SoG cases were modeled to see if there were any differences due to a plus or minus direction of the SoG (Figure 3) on the failure mode and/or durability of the bat.

3.4. Wood Failure Criteria

The three outcomes that are evaluated in the post-processing of the finite element models are (1) no failure, (2) single-piece failure, and (3) multi-piece failure (Figure 7). For each of these modes to be accurately captured by the finite element model, the failure criterion must work in combination with the wood material model to integrate the failure mechanism(s) that initiate and propagate bat breakage with the wood mechanical behavior. The two wood failure options that have been previously explored by the authors are (1) using the failure stress feature in *MAT_WOOD and (2) the *MAT_ADD_EROSION option in LS-DYNA with the strain-to-failure varied as a function of density [6,8,12,19]. The second option was found to correlate well with bat durability testing results and thus is the better choice for the breaking of the bats in the simulations [8,19]. The respective strain-to-failure for the wood density of 0.0245 lb/in^3 (678 Kg/m^3) for each of the wood species in this study is 0.0258 for ash, 0.0236 for maple, and 0.0231 for yellow birch.

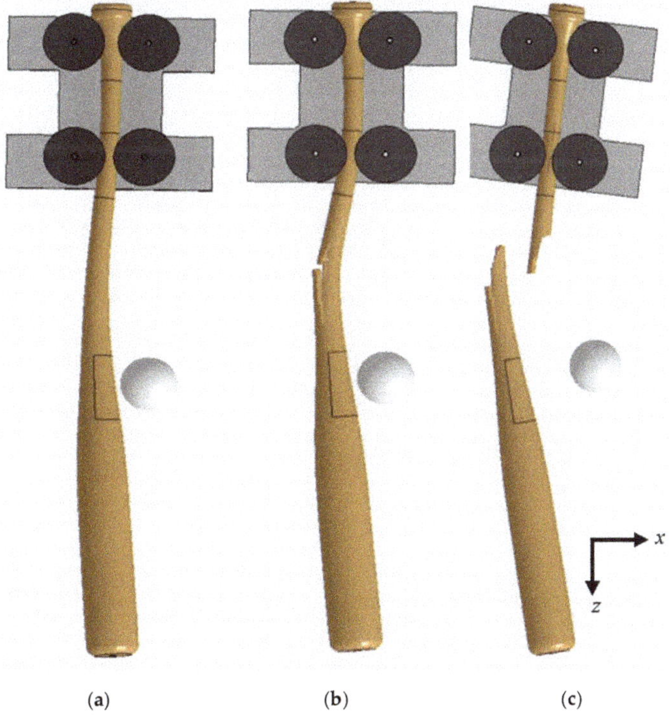

Figure 7. Bat failure modes (**a**) No Failure (NF), (**b**) Single-Piece Failure (SPF), and (**c**) Multi-Piece Failure (MPF).

3.5. Impact Locations

For this study, two impact locations were analyzed, specifically the 2.0 in. (5.1 cm) and 14.0 in. (35.6 cm) locations as measured from the tip of the barrel of the bat (Figure 8). These two impact locations have been observed to be critical locations for bats breaking in MLB games. Impacts at the 2.0 in. (5.1 cm) and 14.0 in. (35.6 cm) locations are away from the nodes associated with the first and second bending modes of vibration, and as a result, induce large amplitudes of vibration along the length of the bat. These high amplitudes translate to large strains, and hence, the potential for bat breakage.

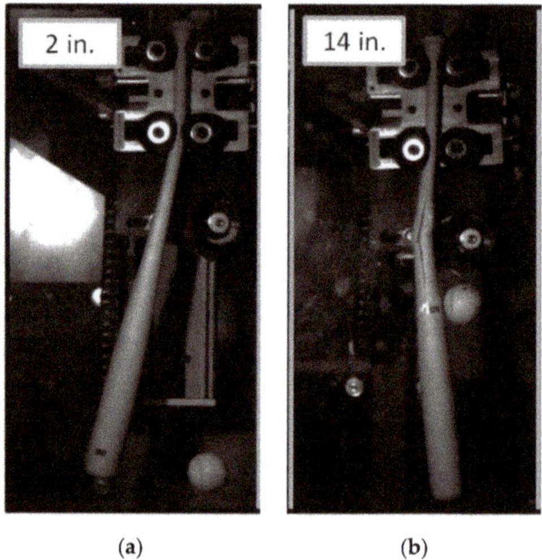

Figure 8. Two common impact locations examined for durability (**a**) Outside pitch—2.0 in (5.1 cm) and (**b**) Inside Pitch—14.0 in. (35.6 cm).

3.6. Impact Velocity

The range of velocities used for this study spanned 90–180 mph (145–290 kph) depending on the impact location of the ball. For the 14.0 in. (35.6 cm) impact model, the velocity range was 90–145 mph (145–233 kph) which corresponds to a 60–100% maximum velocity impact assuming a 90 mph (145 kph) pitch and 90-mph (145 kph) swing speed as measured 2.0 in (5.1 cm) from the tip of the bat rotating about an axis 6.0 in. from the knob to represent the pivot point of the bat in a player's hands during a swing. Using the same methodology, the 2.0 in. (5.1 cm) impact model set had an expanded velocity range up to 180 mph (290 kph) as this velocity corresponds to a maximum 100% velocity impact at the 2.0 in. (5.1 cm) location assuming a 90 mph (145 kph) pitch and 90 mph (145 kph) swing speed. Figure 9 provides a visualization of the impact locations and the respective maximum impact speed.

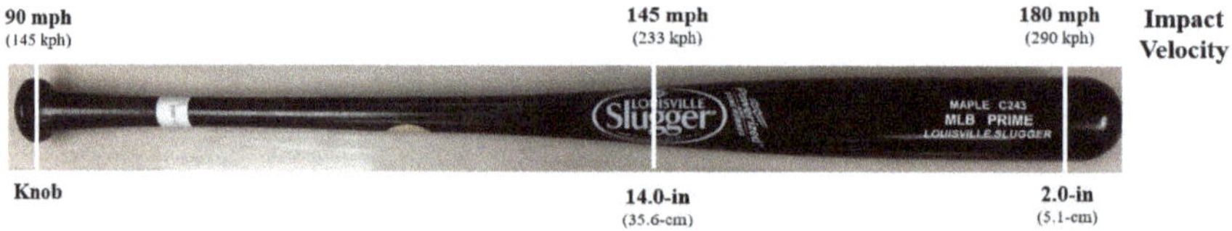

Figure 9. Impact speed variation by impact location.

4. Results

The results of the finite element simulations for a range of wood SoGs were analyzed to investigate the mode of failure and durability as a function of the combination of SoG, impact velocity, wood species, and impact location. For each of the three wood species, 275 simulations were run, i.e., 176 simulations for 2.0 in. (5.1 cm) impact and 99 simulations for 14.0 in. (35.6 cm) impact. In total, 825 bat/ball impact simulations were run for this study.

Contour plots depicting the three outcomes as a function of SoG and impact speed were developed for each combination of wood species and impact location, and the contour plots of the 14.0 in. (35.6 cm) and 2.0 in. (5.1 cm) models are presented in Figures 10 and 11, respectively. The impact-velocity range for each of the possible outcomes of the bats is denoted by color– see the legend in plots. The vertical dark black line in each of these plots denotes the 0° SoG, and the two-vertical dashed black lines bound the range of compliant SoGs as defined by MLB standards for use in baseball bats, i.e., ±3°. In Figures 12 and 13, the results of the three wood species are overlaid for each of the different outcomes. These figures show a direct comparison of the speeds and SoG associated with the respective boundaries between NF to SPF and with the respective boundaries between SPF to MPF.

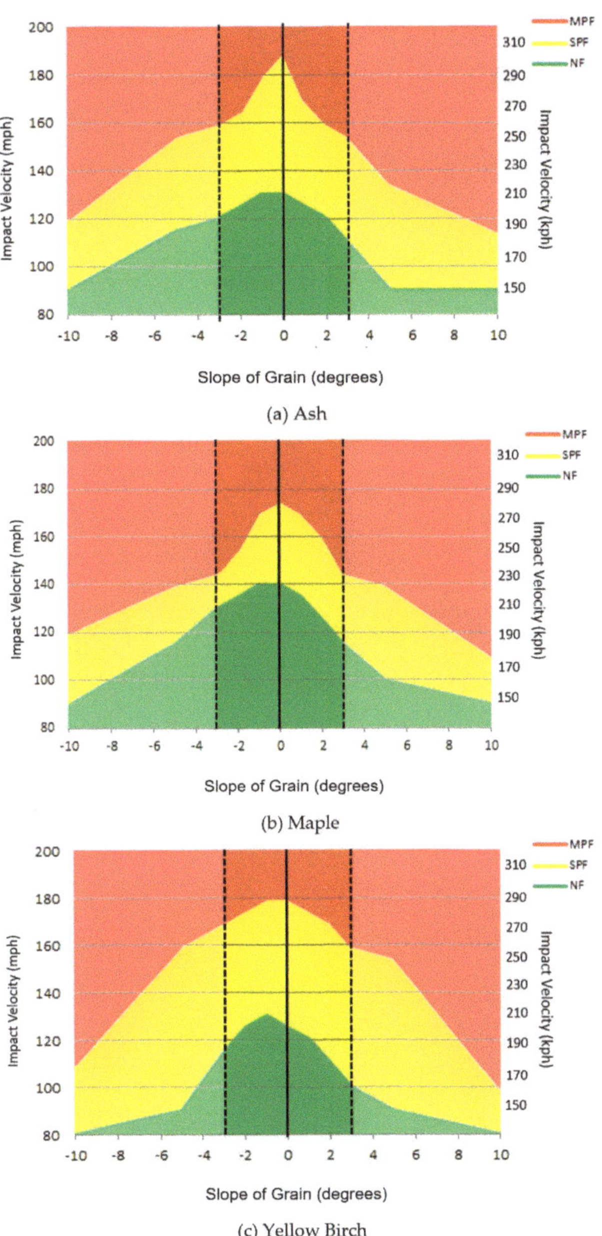

Figure 10. 2 in. (5.1 cm) impact model set results for (**a**) Ash, (**b**) Maple, and (**c**) Yellow Birch.

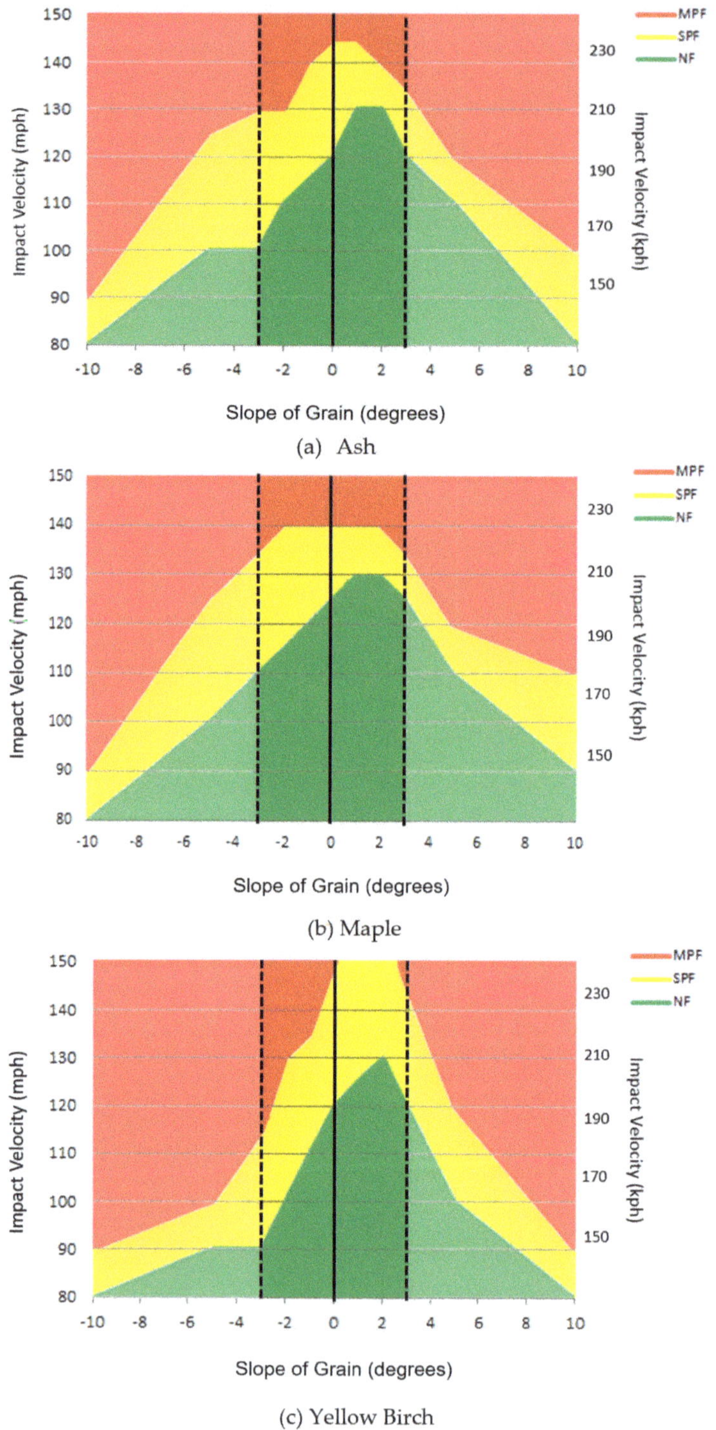

(a) Ash

(b) Maple

(c) Yellow Birch

Figure 11. 14 in. (35.6 cm) impact model set results for (**a**) Ash, (**b**) Maple, and (**c**) Yellow Birch.

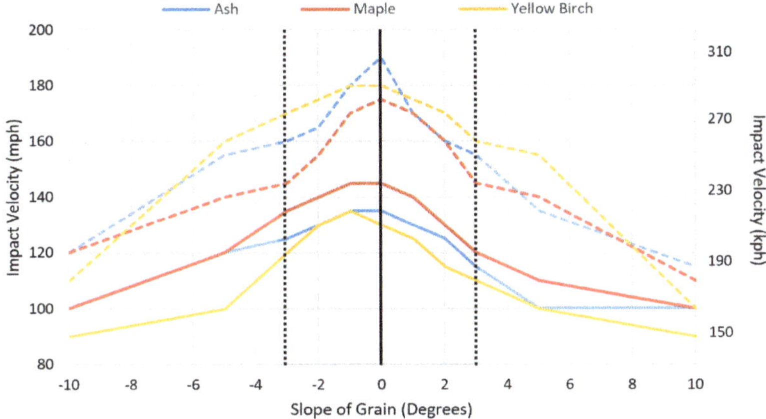

Figure 12. Single-piece (solid lines)/multi-piece failure (dashed lines) threshold for 2 in. (5.1 cm) impact models.

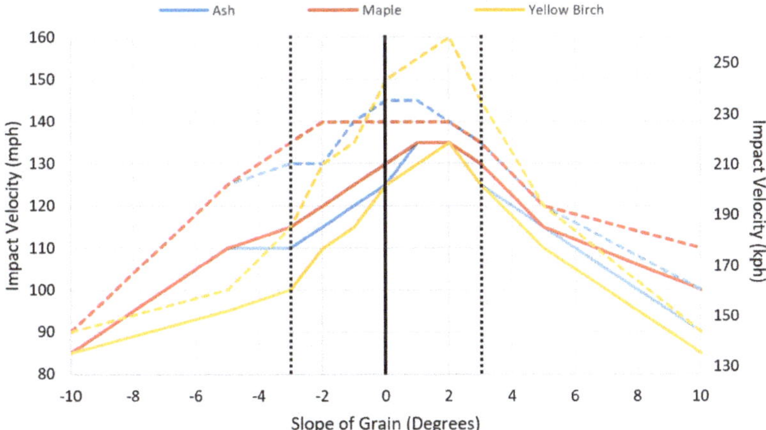

Figure 13. Single-piece (solid lines)/multi-piece failure (dashed lines) threshold for 14 in. (35.6 cm) impact models.

5. Discussion

Figures 10 and 11 are contour plots to show how the response of the bat transitions to each of the three outcomes as a function of impact velocity and SoG for impacts at the 2.0 in. (5.1 cm) and 14.0 in. (35.6 cm) locations, respectively. There is one contour plot for each combination of wood species and impact location that was considered in the parametric study. The NF regions are denoted in green, the SPF regions in yellow, and the MPF regions in red.

Figure 10 shows that the maximum speed for the NF region for all three wood species when impacted at the 2.0 in. (5.1 cm) location is in the range of SoGs progressing from 0° to −1°, and the breaking speed drops for SOGs on either side of this 1° spread. These results of the outside-pitch 2.0 in. (5.1 cm) impact models imply that the negative-SoG wood for values between 0° and −1° can withstand higher impact velocities before failure than the 0°-SoG wood. The relatively narrow bands of SPF for the maple and ash in comparison to the height of the SPF band for yellow birch imply that ash and maple bats quickly transition from SPF to MPF with less increase in speed than for yellow birch bats. It can also be concluded from the contour areas shown in Figure 10 that using wood with an SoG outside of the range of ±3° for any one of the wood species results in a sharp decrease in bat durability. Thus, the 2.0 in. (5.1 cm) modeling provides virtual scientific data to support the 2008 recommendation that the SoG range be set to ±3°.

Figure 11 shows that the maximum speed for the NF region for all three wood species is in the range of SoGs progressing from +1° to +2°. The maximum NF speed then trails off as the SoG increases to +3° and higher and as the SoG goes from +1° and lower. These results of the 14.0 in. (35.6 cm) inside-pitch impact models show that the positive-SoG wood for values between 0° and 3° can withstand higher impact velocities before transitioning to an SPF than the negative-SoG wood. It can also be concluded from the data shown in Figure 11 (as it was from Figure 10) that using wood with a SoG outside of the range of ±3° for any one of the wood species results in a sharp decrease in bat durability, which also supports the recommendation that the SoG range be set to ±3°.

Figure 12 is an overlay plot of the boundaries denoting the transition from NF to SPF (solid lines) and for the transition from SPF to MPF (dashed lines), respectively, for all three wood species when impacted at the 2.0-in. (5.1-cm) location. Figure 12 shows that the yellow birch bats transition from NF to SPF at lower velocities for the entire range of ±10° SoGs in comparison to the maple bats and for most of the range of ±10° SoGs relative to the ash bats. However, Figure 12 also shows that the yellow birch bats had a much higher durability for the transition from SPF to MPF than maple and slightly better than ash. Thus, maple is the most likely of the three wood species to result in an MPF at the high end of the bat/ball impact speeds when viewing the results in the SoG range of ±3°, and this probability of maple breaking in an MPF manner relative to ash and yellow birch increases for SoGs outside the ±2° range. Thus, the modeling implies that the rate of maple bat MPFs due to inside pitches could be measurably reduced if the SoG range for maple were decreased from ±3° to ±2°.

Figure 13 is an overlay plot of the boundaries denoting the transition from NF to SPF (solid lines) and for the transition from SPF to MPF (dashed lines), respectively, for all three wood species when impacted at the 14.0 in. (35.6 cm) location. The yellow birch bats transition from NF to SPF at much lower velocities in comparison to the maple bats and slightly lower than the ash bats. For example, the transition velocities from NF to SPF for a bat with a −3° SoG for maple and ash are 115 and 110 mph, respectively, while the yellow birch is 100 mph. However, the yellow birch bats once again had a much better durability window (for both velocity and SoG span) in their transition from SPF to MPF than the other two species. This 14.0-in. (35.6-cm) modeling implies that the rate of MPFs due to outside pitches could be reduced if the SoG range for all three kinds of wood were decreased from ±3° to ±2°.

There are several general observations that can be made from the analysis of the models. Negative-SoG bats in the range of 0° to −1° exhibit a higher SPF durability than positive-SoG bats for all three wood species for a 2.0 in. (5.1 cm) impact. As the impact speed increases to the MPF boundary, the best durability is at a 0° SoG. Positive-SoG bats in the range of +1° to +2° exhibit a higher SPF durability than negative-SoG bats for all three wood species for a 14.0 in. (35.6 cm) impact. This phenomenon continues for yellow birch bats as they transition from an SPF to an MPF. As the impact speed increases to the MPF boundary, the best durability is at a 0° SoG for ash and maple bats. The bias of the peak durabilities to be for negative-SoGs for outside-pitch (2.0 in., 5.1 cm) impacts and to be for positive SoGs for the inside-pitch (14.0 in. 35.6 cm) impacts supports that a 0°-SoG bat is the most versatile with respect to bat durability.

Based on the Hankinson equations as plotted in Figure 4, it was expected that bat durability would decrease as the magnitude of SoG increases. Thus, it is unexpected to see the results of increased bat durability for small nonzero SoGs, i.e., increase in bat durability for small negative SoGs for the 2.0 in. (5.1 cm) outside-pitch impact and the increase in bat durability for small positive SoGs for the 14.0 in (35.6 cm) inside-pitch impact. The respective increases with opposite SoGs may be a consequence of the respective deformation mode and location of the maximum strain for each of the two impact conditions. For the outside-pitch condition, the bat exhibits a diving board mode shape, i.e., like that of a beam clamped at one end with a backward tip deflection—see Figure 8a. For the inside-pitch condition, the bat bends in the reverse direction of that described for the outside-pitch situation with the zero-slope near the middle of the length of the bat—see Figure 8b. Thus, for the outside pitch, the front side of the bat is in tension with the maximum strain in the handle section near the knob. For the inside pitch, the back side of the bat is in tension with the maximum strain developing in the taper section of the bat. This bat response was visually confirmed during high-speed video recordings of impacts and could be validated experimentally through digital image correlation (DIC) or by placing strain gauges along the length of the bat. The difference in the sensitivity of the range of SoG may be a consequence of the respective bat section that is experiencing the maximum strain for the respective impact location. With respect to how the material behavior contributes to this increase in bat durability for small non-zero SoGs, a plausible explanation may be at what magnitude of bending do the wood fibers become fully engaged in the deformation process. Future studies will look at extreme changes in bat profiles to see if the trends found for this bat profile are also seen in those other profiles. Simple parametric studies of beam bending for the range of SoGs used in this parametric study may give insight into the specifics of the mechanical behavior of the material that may explain the results.

There are several minor limitations with the model used in this paper. The SoG is modeled as the same throughout the length of the bat. In reality, there will be some variation of the SoG along the length of the bat. However, this limitation is minor because the SoG in the handle dominates over the SoG elsewhere on the bat. There are only a few lab experimental results to use for comparison to the model. The wood properties are taken as nominal.

6. Conclusions

Finite element models of the popular C243 bat profile, baseball, and a bat durability system were constructed and used to study the effect of SoG on the outcomes of ash, maple, and yellow birch bats for inside and outside pitches. For an inside-pitch impact, bats with a positive SoG exhibited better durability than negative-SoG bats—with the best durability for bats with SoGs in the range of +1° to +2°. Conversely, negative-SoGs performed best for outside-pitch impacts when compared to results of positive-SoG bats of the same profile and material properties. For both impact locations, sharp drop-offs in performance were observed as the SoG went outside the range of ±3°. When comparing the three wood species, the yellow birch bats failed at the lowest velocity thresholds but exhibited the highest velocity range of the single-piece failure. The maple bats exhibited the highest velocity threshold before a single-piece failure occurred but were the most likely to fail multi-piece. The ash bats exhibited a balance of moderate velocity threshold before failure with a smaller range of multi-piece failure.

Author Contributions: J.F.-S.'s contributions to the research were formal analysis of the finite element modeling studies presented in the article, investigation of the effect slope of grain has on the durability of baseball bats comprised of three different wood species, development of a methodology to study the effect of slope of grain on wood baseball bat durability through finite element modeling, validation and visualization of the finite element results, writing the original draft and reviewing and editing the paper. J.S.'s author contributions to this work were conceptualization of the project goals, funding acquisition for the project, aiding in the development of the methodology used accomplish the finite element study, acting as a project administrator and supervisor to the project, and reviewing and editing the writing of the article. P.D.'s author contributions to the article are conceptualization of the project goals, formal analysis of the finite element modeling results, project administration and supervision of the research goals and tasks to complete the project, providing resources to properly evaluate the finite element modeling results, and reviewing and editing the article. E.R.'s author contributions to the research were providing finite element modeling resources and developing the finite element model methodology that served as a guideline to the parametric studies presented in the article. B.C.'s author contributions to the article were reviewing and editing the paper. D.K.'s author contributions to the article were funding acquisition for the research, serving as a project administrator and supervisor to the project goals, and serving as a crucial resource in regard to wood science and wood behavior.

Funding: This research was funded in part by the Office of the Commissioner of Baseball and the MLB Players Association.

Acknowledgments: This material is based upon the work supported by the Office of the Commissioner of Baseball and the MLB Players Association. Any opinions, findings, and conclusions or recommendations expressed in this material are those of the authors and do not necessarily reflect the views of the Office of the Commissioner of Baseball and the MLB Players Association.

Conflicts of Interest: The authors declare no conflict of interest. The funding sponsors had no role in the design of the study; in the collection, analyses, or interpretation of data; in the writing of the manuscript, and in the decision to publish the results.

References

1. Jensen, D. *The Timeline History of Baseball*; Thunder Bay Press: San Diego, CA, USA, 2009.
2. Cannella, S. *Against The Grain Carving a New Niche in Maple, A Canadian Woodworker Gave Hitters an Alternative to Ash and Unleashed the Boutique Bat Business*; Sports Illustrated: New York, NY, USA, 2002.
3. MLB/MLBPA MLB. *MLBPA Adopt Recommendations of Safety and Health Advisory Committee December*; Major League Baseball: New York, NY, USA, 2008.
4. Dvorchak, R. Bat Maker a big hit with the pros, SPORTS P. A-1. *Pittsburgh Post-Gazette*, 4 July 2010.
5. Saraceno, J. Taking swing at safer bats, man says his creation curbs exploding effect. *USA TODAY*, FA CHASE EDITION P.1C. 15 August 2016.
6. Fortin-Smith, J.; Sherwood, J.; Drane, P.; Kretschmann, D. Characterization of Maple and Ash Material Properties for the Finite Element Modeling of Wood Baseball Bats. *Appl. Sci.* **2018**, *8*, 2256. [CrossRef]
7. Ruggiero, E.; Sherwood, J.; Drane, P.; Kretschmann, D. An Investigation of bat durability by wood species. In Proceedings of the ISEA 2012, 9th Conference of the International Sports Engineering Association, Lowell, MA, USA, 9–13 July 2012.
8. Ruggiero, E.; Sherwood, J.; Drane, P. Breaking Bad(ly)—Investigation of the Durability of Wood Bats in Major League Baseball using LS-DYNA®. In Proceedings of the 13th LS-DYNA International Users Conference, Dearborn, MI, USA, 8–10 June 2014.
9. Carrasco, E.V.M.; Mantilla, J.N.R. Applying Failure Criteria to Shear Strength Evaluation of Bonded Joints According to Grain Slope under Compressive Load. *Int. J. Eng. Technol.* **2013**, *13*, 19–25.
10. Hankinson, R.L. *Investigation of Crushing Strength of Spruce at Varying Angles of Grain*; Air Force Information Circular No. 259; The Chief of Air Service: Washington, DC, USA, 1921.
11. Hernandez, R. About RockBats 2012. Available online: http//:www.rockbats.com/company_aboutRockBats.html (accessed on 8 April 2018).
12. Kretschmann, D.E.; Bridwell, J.J.; Nelson, T.C. Effect of changing slope of grain on ash, maple, and yellow birch in bending strength. In Proceedings of the WCTE 2010, World Conference on Timber Engineering, Riva del Garda, Trento, Italy, 20–24 June 2010.
13. Glier, R. Wood experts will study bats further. *USA TODAY*, SPORTS P. 5C. 4 August 2009.
14. Drane, P.; Sherwood, J.; Shaw, R. An Experimental Investigation of Baseball Bat Durability. In *The Engineering of Sport 6*; Moritz, E.F., Haake, S., Eds.; Springer: New York, NY, USA, 2006.
15. Ruggiero, E.; Fortin-Smith, J.; Sherwood, J.; Drane, P.; Kretschmann, D. Development of a Protocol for Certification of New Wood Species for Making Baseball Bats. *J. Sports Eng. Technol.* **2017**. submitted for review.
16. Ruggiero, E. Investigating Baseball Bat Durability Using Experimental and LS-DYNA Finite Element Modeling Methods. Master's Thesis, University of Massachusetts Lowell, Lowell, MA, USA, 2011.
17. Murray, D.; Reid, J.D.; Faller, R.K.; Bielenberg, B.W.; Paulsen, T.J. *Evaluation of LS-DYNA Wood Material Model 143*; PUBLICATION NO. FHWA-HRT-04-096; Federal Highway Administration: McLean, VA, USA, 2005.
18. *Manual for LS-DYNA Wood Material Model 143*; U.S. Department of Transportation Federal Highway Administration: Washington, DC, USA, 2007.
19. Drane, P.; Sherwood, J.; Colosimo, R.; Kretschmann, D. A study of wood baseball bat breakage. In Proceedings of the ISEA 2012, 9th Conference of the International Sports Engineering Association, Lowell, MA, USA, 9–13 July 2012.

Article

Mechanical Characterisation and Modelling of Elastomeric Shockpads

David Cole [1,*], Steph Forrester [1] and Paul Fleming [2]

[1] Wolfson School of Mechanical, Electrical and Manufacturing Engineering, Loughborough University, Loughborough LE11 3TU, UK; S.Forrester@lboro.ac.uk
[2] School of Architecture, Building and Civil Engineering, Loughborough University, Loughborough LE11 3TU, UK; P.R.Fleming@lboro.ac.uk
* Correspondence: D.C.Cole@lboro.ac.uk; Tel.: +44-1509-564-827

Received: 30 January 2018; Accepted: 20 March 2018; Published: 27 March 2018

Abstract: Third generation artificial turf systems are comprised of a range of polymeric and elastomeric materials that exhibit non-linear and strain rate dependent behaviours under the complex loads applied from players and equipment. An elastomeric shockpad is often included beneath the carpet layer to aid in the absorption of impact forces. The purpose of this study was to characterise the behaviour of two elastomeric shockpads and find a suitable material model to represent them in finite element simulations. To characterise the behaviour of the shockpads an Advanced Artificial Athlete test device was used to gather stress-strain data from different drop heights (15, 35 and 55 mm). The experimental results from both shockpads showed a hyperelastic material response with viscoelasticity. Microfoam material models were found to describe the material behaviour of the shockpads and were calibrated using the 55 mm drop height experimental data. The material model for each shockpad was verified through finite element simulations of the Advanced Artificial Athlete impact from different drop heights (35 and 15 mm). Finite element model accuracy was assessed through the comparison of a series of key variables including shock absorption, energy restitution, vertical deformation and contact time. Both shockpad models produced results with a mean error of less than 10% compared to experimental data.

Keywords: shockpad; artificial turf; rubber; finite element analysis; impact

1. Introduction

Artificial turf has become an increasingly prominent type of playing surface for many sports due to its versatility for all-year-round and multi-sport use [1]. Whilst natural turf remains the star quality for many sports, the advantages of artificial surfaces over natural pitches means their use, especially at the grassroots level, is increasing [2]. The Fédération Internationale de Football Association (FIFA) have described the latest 3G surfaces as "the best alternative to natural grass" due to their resistance to weather and ability to sustain more intensive use [3]. All these factors led the English football association to commit to an investment of £230 million for new 3G pitches by 2020 [4].

Third generation (3G) artificial surfaces contain a number of polymer materials that exhibit various strain rate and temperature dependant behaviours [5]. The surface is constructed in a layered format [6] consisting of a lower layer of an elastomeric mat called a shockpad, used to absorb the impact forces of boot-surface interactions and thereby reduce the stress upon the athletes (Figure 1). On top of this lies the carpet consisting of a bed of polymer fibres arranged in tufts and stitched into a canvas backing. The carpet is filled with two separate infill layers: a stabilising layer to provide support to the fibres and weigh the carpet down and a performance layer to fill the carpet fibres and provide an interface for ball and player interactions. The stabilising layer is typically created using sand whilst the performance layer can consist of a variety of different polymer or organic materials however styrene butadiene rubber (SBR) granules are the most common choice.

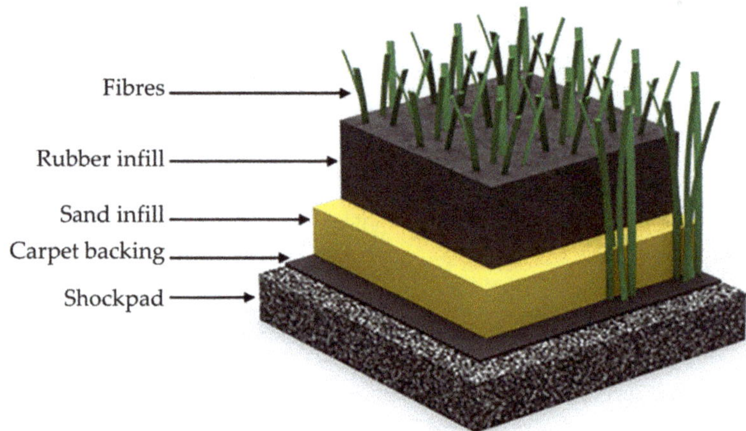

Figure 1. Layered construction of 3G turf surface.

The main design requirement of the shockpad layer is to absorb some of the impact forces applied from player interactions such as running foot strikes. They must also allow for drainage and be durable to withstand cold temperatures and repeated freeze-thaw cycles [7]. Elastomers are a popular choice as they have favourable shock absorption properties and can deform to large strains without permanent deformation. These properties result from the chemical structure of elastomers being made up of long randomly oriented molecular chains that align when stretched and restore when the load is released. The term elastomer, a combination of elastic and polymer is often used interchangeably with rubber [8]. Rubber is a popular choice of elastomer for engineering applications due to its flexibility, extensibility, resiliency and durability [9]. Shockpads are typically constructed to be cellular, a move that enhances energy absorption properties and allows drainage through the system [10]. The presence of air voids also allows larger strains to be reached under compression [11]. Elastomer foams can be found in many sports products, a common example being in sports footwear acting as shock absorbers [12]. A key difference between these foams however is their construction method. Whilst most elastomeric foams are formed with the use of a blowing agent, shockpads are typically formed using recycled rubber compounds bound together with adhesive with the voids in the structure formed by the spaces between solid rubber particles [6].

The behaviour of shockpads under compression has been analysed previously. Both Anderson [13] and Allgeuer et al. [14] identified three key stages: air void compression, transition and rubber compression. The air void compression phase is associated with an initial low stiffness, high deflection response. A transition phase then provides an increase in the stiffness as the void ratio is reduced and the shockpad turns into a two phase system of rubber and binder. A final phase is realised when rubber-on-rubber contact is made resulting in a substantial increase in stiffness due to the compression of the rubber.

The modelling of sports surfaces has been attempted before with examples of running tracks, vinyl flooring and rubber treadmill belts all investigated [15–18]. Anderson [13] created a mechanical model to describe the three regions in shockpad compression using a non-linear damped model consisting of a linear spring and damper in parallel, similar in nature to a hyperelastic Bergstrom-Boyce model [19]. Mehravar et al. [20] used a finite element (FE) model to simulate the loading of an elastomeric shockpad using an Arruda-Boyce hyperelastic material model. Compression data was used to fit the coefficients of the model and a power equation created to describe the frequency dependency of the strain energy density function. The simulation was able to accurately fit the loading response under compression for frequencies in the range 0.9–10 Hz. However, the model was restricted to describing the loading behaviour with the viscoelastic unloading response not considered. The purpose of this study was to mechanically characterise the behaviour of two elastomeric shockpads and to find suitable material models to describe both loading and unloading behaviours. The mechanical behaviour of the rubber shockpads were characterised through calculation of the stress-strain properties obtained from testing using the FIFA standard Advanced Artificial Athlete (AAA, FIFA, 2015).

2. Shockpad Characterisation

The process of developing a material model relies on understanding the behaviour of the material that is to be modelled. Calibration of a material model requires optimising model parameters against experimental data collected under similar loading conditions. As shockpads primarily deal with compressive loads, data was recorded using a AAA impact tester [21]. The AAA is a device used to measure the shock absorption, energy restitution and vertical deformation properties of a surface and thereby assess the standard of the surface [22]. Stress-strain data from a standard drop height of 55 mm was used to develop the material model and experimental data from additional drop heights (15 and 35 mm) used to validate the material model in a series of FE simulations.

2.1. Shockpad Properties

Two shockpads were chosen for analysis, one prefabricated and another that was laid in situ on Loughborough University campus. Both shockpads were FIFA quality approved and were comprised of rubber aggregate bound by polyurethane adhesive (Table 1). The prefabricated shockpad was Regupol® 6010 SP (BSW Berleburger GmbH, Berleburger [23]) designed specifically for 3G turf surfaces (Figure 2a). The in-situ shockpad was created during the construction of a 3G turf sports pitch at Loughborough University in 2014 [6] (Figure 2b). The in-situ shockpad samples used in lab testing were laid onto a 10 mm thick plywood board during the construction of the surface and were subsequently transferred to the rigid lab floor and cut to size. As the shockpads were both cellular materials, relative density was calculated. Relative density is often used as a way of expressing cellular materials with respect to the 100% dense solid [24]. It is calculated by dividing the density of the cellular material by the density of the solid. As the exact rubber density for both shockpads was not reported, a value of 1100 kg/m^3 was assumed [25].

(a)   (b)

Figure 2. Sample sections of (**a**) Regupol® 6010 SP shockpad manufactured by Berleburger. (**b**) Holywell shockpad laid in-situ on Loughborough University campus.

Table 1. Shockpad properties overview.

Name	Material	Granule Size (mm)	Binder	Density (kg/m^3)	Thickness (mm)	Relative Density
Berleburger	Rubber shreds	1–2	Polyurethane	557	15.0 ± 0.1	0.49
Holywell	Rubber granules	2–6	Polyurethane	575–600	22.4 ± 0.9	0.54

2.2. AAA Methodology

The AAA test device is designed to measure and record the shock absorption and deformation behaviour of 3G turf surfaces under human impact conditions to assess whether they are suitable for sporting use [21]. The device consists of a convex test foot mounted to a 20 kg mass via a spiral steel spring (Figure 3). A remote controlled electromagnet releases the mass from a height of 55 mm and the convex test foot with diameter 70 mm acts as an impact face with the surface. An accelerometer is attached to the base of the falling mass and records the acceleration throughout the drop, impact and rebound. Shock absorption, energy restitution and vertical deformation are all calculated from the accelerometer output for each drop.

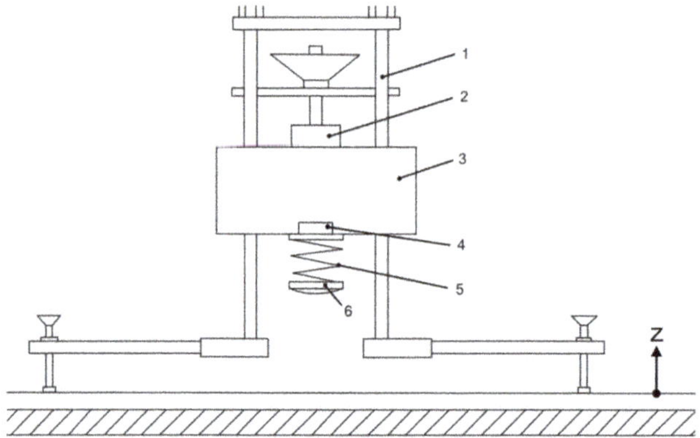

Figure 3. Advanced Artificial Athlete (AAA) schematic: 1. support frame; 2. electric magnet; 3. 20 kg falling mass; 4. accelerometer; 5. linear stiffness spring; 6. 70 mm diameter test foot [21].

Shockpad samples of at least 30 × 30 cm were laid beneath the impact foot and were subjected to drops from multiple heights from 55 mm down to 5 mm in 10 mm increment steps. Three drops were used for each drop height as per the FIFA test standard [21]. The different drop heights allowed the response of the shockpads under different impact energies to be examined. Raw accelerometer data was extracted from each of the test drops and processed using Matlab (Version 2017a, Mathworks, Natick, MA, USA). Velocity and displacement were calculated through integration using the trapezium rule for the duration of each drop. Shock absorption, energy restitution and vertical deformation were calculated from the first drop at each drop height using methods set out in the FIFA handbook of test methods [21] with the exception of the shock absorption which was altered to ensure the reference value for the solid floor was adjusted for each drop height. Contact time was taken as the time between the minimum vertical velocity (z-axis; Figure 3) that occurred as the impactor first contacted the surface, and the maximum vertical velocity which occurred as the impactor left the surface. The force and displacement data, obtained from the acceleration as detailed above, were converted into engineering stress and strain. Curvature of the test foot was neglected within the engineering stress calculations; a fixed cross-sectional area (of 38.5 cm^2) was used throughout contact. Unloaded shockpad thickness, the average over 15 measurements (Table 1), was used to calculate the engineering strain.

2.3. AAA Experimental Results

For both shockpads the stress-strain results were highly consistent between the three drops performed from each height (Figure 4). The loading phase of the Berleburger shockpad demonstrated an undulating loading response consisting of a series of higher stiffness regions followed by a plateau in the stress. In contrast, the Holywell shockpad showed a more typical hyperelastic loading response. Both shockpads also demonstrated hysteresis during the unloading phase. The Berleburger shockpad was the stiffer of the two shockpads and had a higher maximum stress. The undulations during the loading phase in the Berleburger shockpad were not as evident for the lower drop heights, while the Holywell shockpad followed a similar loading response for all drop heights (Figure 5). As the drop height decreased the shockpads showed similar responses albeit with a lower maximum stress. At the lowest two drop heights of 5 and 15 mm the stress-strain response was not large enough to enter into a hyperelastic region and remained relatively linear. Shock absorption increased as the drop height was lowered in both shockpads. Strain levels were similar for both the shockpads due to the difference in thickness. Both shockpads remained in their elastic region and since they produced similar results for all three drops from a given height, only the first drop was used for the purpose of material modelling and simulation.

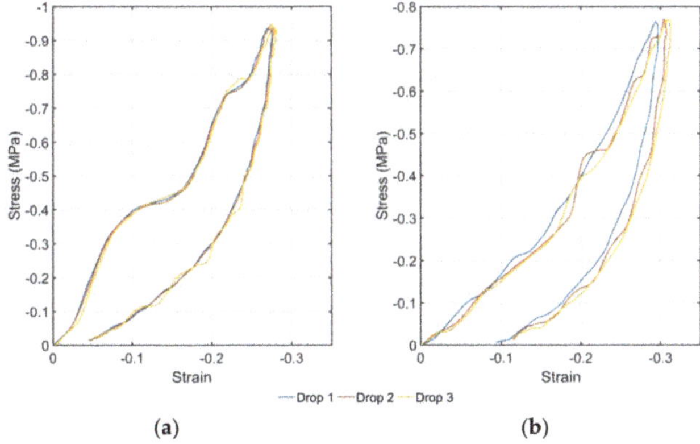

Figure 4. Stress-strain curves from 3 drops at the 55 mm drop height for the (**a**) Berleburger shockpad and (**b**) Holywell shockpad.

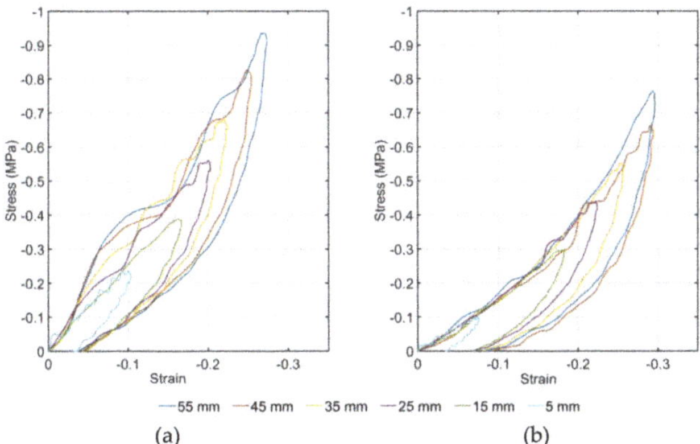

Figure 5. Stress-strain curves for the first drop for each drop height for (**a**) Berleburger and (**b**) Holywell shockpad.

3. Finite Element Simulation of AAA

A FE model of the experimental AAA test on the shockpads was created in Abaqus CAE (Figure 6). The purpose of the model was to try and replicate the shockpad response from different height drops after calibration of a material model to a single drop height. The shockpad layer was modelled as a deformable part and was placed on top of a fixed rigid concrete block. The shockpad was meshed using 3D 8-node reduced integration elements with hourglass control (C3D8R) with an increased mesh density around the impact area. A mesh sensitivity test was completed to ensure results were not sensitive to mesh design. A tangential friction was specified between the shockpad and concrete using a penalty formulation with a friction coefficient of 0.8. This value was measured as the coefficient of sliding friction in an experimental slip test performed prior to AAA testing. The AAA was modelled as a rigid steel test foot with point mass placed directly above the centre point. The test foot and point mass were joined by a deformable spring element with a 2000 N/mm stiffness as per the experimental test [21]. All parts of the AAA were restrained to only allow translation in the z-axis. Starting conditions for the test placed the impact foot 1 mm above the shockpad top surface with an initial velocity matching those seen in the experimental tests. Acceleration, velocity and displacement of the point mass were outputted from the simulation and used to calculate the stress-strain response as well as the FIFA test outputs. Similarly, to the stress-strain calculations for the experimental data, engineering stress was based on a fixed cross-sectional area throughout contact corresponding to full test foot contact area (38.5 cm^2). Due to the symmetrical nature of the problem only a quarter portion of the model was constructed to save on computing time.

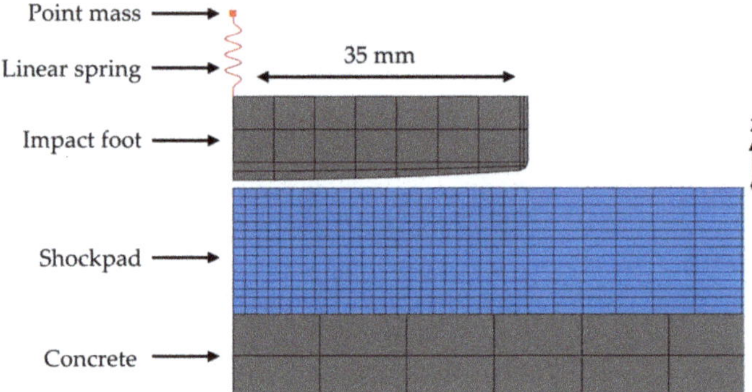

Figure 6. AAA FE model showing increased mesh density around impact area. The model was extruded to eight times the radius of the impact foot.

Material Model Development

Development of material models to represent the behaviour of the shockpads followed a systematic approach (Figure 7). From the experimental results, it was clear that hyperelastic material models were needed to match the non-linear deformation behaviour of the shockpads (Figure 5). Hysteresis was also seen for both shockpads meaning models capable of capturing viscoelasticity must be used. Taking these factors into account, two models were selected for initial calibration. The Bergstrom-Boyce model has been successful in representing solid rubber behaviour [19], however, due to the cellular nature of the shockpads a microfoam material model was also selected. Separate models were calibrated for each of the shockpads in order to capture their unique behaviour. MCalibration software (Version 4.6, Veryst Engineering, Needham, MA, USA) was used to calibrate models from the PolyUMod material database against stress-strain data from a single AAA drop from 55 mm for each of the shockpads (Table 2). The MCalibration software provides a more extensive library of material models compared to the inbuilt Abaqus materials library and gives the user more control over the calibration. Each of the material coefficients were calibrated to the experimental data using equations including Lavenberg-Marquardt, NEWUOA and Nelder-Mead. Whilst this process fits the model to the data, the software does not understand relationships between variables often meaning further manual alterations must be done. Calibrated material models were imported into the FE simulation where the AAA drop from 55 mm was replicated and the results of the simulation subsequently compared to the experimental data. A number of variables were used as an indicator of the model accuracy, including the shock absorption, energy restitution, vertical deformation and contact time. The root mean square error (RMSE) was also calculated for the strain as the AAA impact was a stress controlled test. The model was deemed accurate if the mean error was less than 10% and no individual error was greater than 15%. If these criteria were not met the model coefficients were refined manually in MCalibration. Error was defined as the percentage difference between the experimental values and the FE simulated values with exception of the strain RMSE that was expressed as a percentage of the maximum strain. Refinement of the models was done using an intelligent trial and error approach, changing the appropriate coefficients to alter the material response based upon the results of the previous simulation. Both models calibrated well to the experimental data, however, the Bergstrom-Boyce model was unable to accurately represent the deformation response to an error lower than 15%. The microfoam material model performed much better, producing good initial results as the relative density was taken into consideration during the model calculations. This model was further refined to increase the accuracy for both shockpads before validation. Validation of the models was undertaken through the simulation of the lower drop heights of the AAA, with the results subsequently compared to the experimental data.

Table 2. Optimised microfoam model coefficients for the two shockpad models. Stress outputted in Pascals.

Microfoam Model Properties			Shockpad Model Values	
Parameter Name	Unit	Description	Berleburger	Holywell
Es	Stress	Young's modulus when no porosity	7,000,000	5,500,000
alphaE	-	Modulus density scaling factor	0.0056	0.0056
hE	-	Modulus density scaling factor	2.5	2.5
nu0	-	Poisson's ratio in the limit of 100% porosity	0.05	0.05
nus	-	Poisson's ratio in limit of no porosity	0.49	0.2
rhor	-	Reduced density of the material	0.5	0.54
lambdaL	-	Limiting chain stretch	4	4
sB	-	Relative stiffness of network B	5	5
p0	Stress	Initial gas pressure inside the foam voids	0	0
xi	-	Strain adjustment factor	0.05	0.05
C	-	Strain exponential	-0.5	-0.5
tauHat	Stress	Normalised flow resistance	300,000	300,000
m	-	Stress exponent	2	2

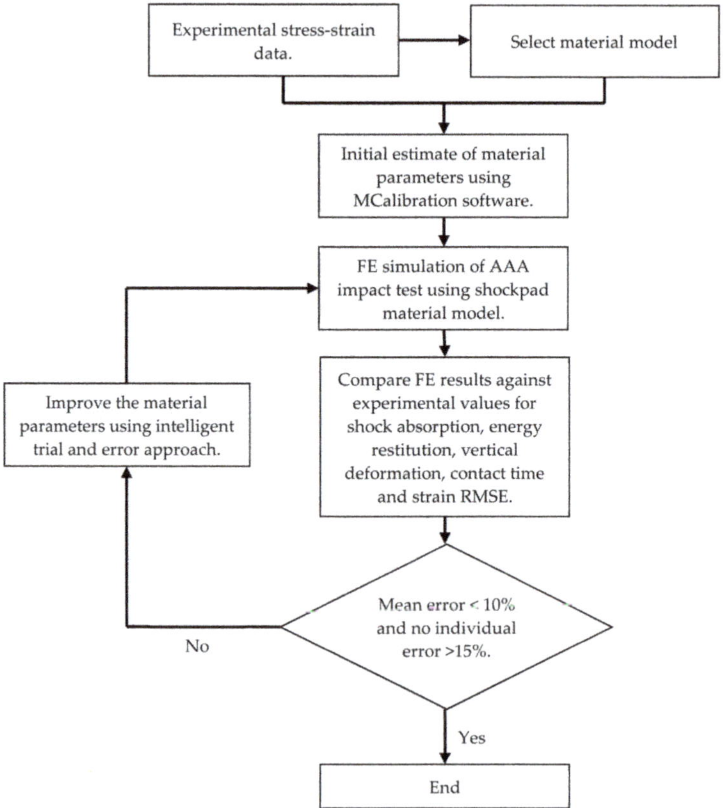

Figure 7. Flow chart showing the steps involved in material model development.

4. Results

The microfoam material model showed good overall agreement with the experimental data for both the Berleburger and Holywell shockpads and all drop heights (Figure 8). The FE simulations met the accuracy criteria for both shockpads and all drop heights, except for the Berleburger 15 mm drop height where the strain RMSE was slightly above the threshold at 18% (Table 3).

The FE simulations under-estimated shock absorption for the Berleburger shockpad by a consistent 3–4% across drop heights, corresponding to an over-estimation in peak force. Agreement was better for the Holywell shockpad with the difference within 2% throughout. The FE simulations over-estimated vertical deformation for the Berleburger shockpad, by up to 0.3 mm, while agreement was again better for the Holywell shockpad with the difference being no more than 0.2 mm throughout. In all cases if shock absorption was under-estimated then vertical deformation was over-estimated and vice versa, suggesting that for both shockpads the overall loading stiffness response was better estimated than the two loading extreme measures of shock absorption and vertical deformation. The FE simulations over-estimated energy restitution by up to 8% suggesting that more energy was returned to the impactor than was observed experimentally. These differences were larger than those for shock absorption and vertical deformation which was not unexpected given that both the loading and unloading phases contributed to this variable [20]. Interestingly, for both shockpads, the difference in energy restitution decreased with decreasing drop height; from 6 to 8% for the 55 mm drop height to only 0–1% for the 15 mm drop height. This trend suggests that the loading–unloading response is accurately captured by the simulations for low strains, but less so for higher strains that reach the hyperelastic region of the stress-strain curve (Figure 5). For low strains, the compression and recovery are dominated by air voids, while for higher strains the compression and recovery includes not only air voids but also the rubber shreds/granules. This poorer performance may, therefore, represent a limitation of using a microfoam model to estimate compression and recovery of a granular/shredded rubber held together by a polyurethane binder (Table 1).

Strain RMSEs were also larger than those for shock absorption and vertical deformation, on average similar between shockpads, but showed no consistent trend with drop height. For the Berleburger shockpad the strain RMSE increased with decreasing drop height (from 6 to 18%) and appeared linked to an increasing offset in the unloading curves between the experimental data and simulation output. In contrast, for the Holywell shockpad the strain RMSE decreased with decreasing drop height (from 14% to 6%) and appeared to be linked to the simulation output exhibiting undulations during the loading phase that, as noted above, were not present in the experimental data for this shockpad, with the undulations being of greater magnitude for the higher drop heights (Figure 8).

Although some of these differences will be due to inaccuracies in the material models and FE simulations, there will also be a contribution related to uncertainties in the experimental data itself. These include: inconsistency in the shockpad thicknesses (Table 1) due to construction method and granular nature; errors introduced through numerical integration of the accelerometer data to generate the velocity and displacement data [20]; and in the use of this calculated velocity data to identify the instant of impact and impact conditions.

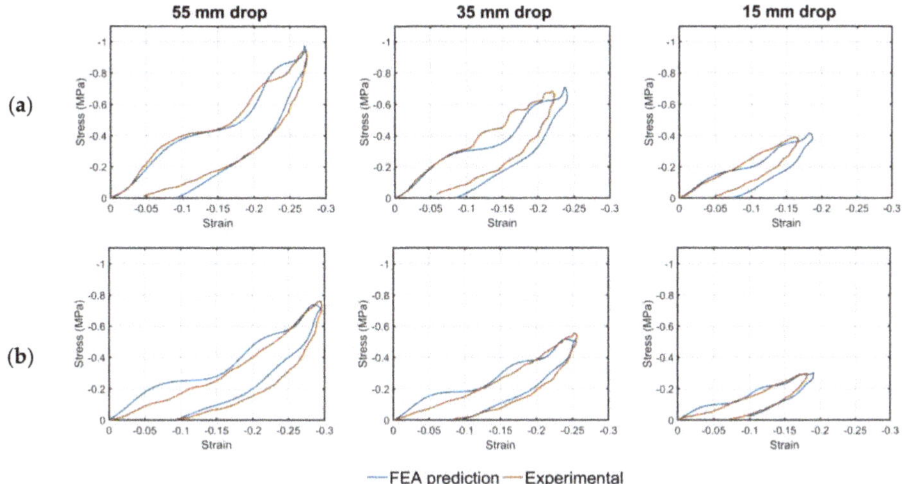

Figure 8. Comparison of stress-strain experimental data against FE simulation outputs across the three drop heights for (**a**) Berleburger and (**b**) Holywell shockpads.

Table 3. Experimental and FE simulation results from first drop of AAA for the Berleburger and Holywell shockpads.

Shockpad and Drop Height	Source	Shock Absorption (%)	Energy Restitution (%)	Vertical Deformation (mm)	Contact Time (ms)	RMSE (% of Max Strain)	Mean Error (%)
Berleburger 55 mm	Experimental	41	55	−4.1	18.5	-	-
	FE	38	63	−4.1	17.4	-	-
	Difference (%)	8	13	0	6	6	7
Berleburger 35 mm	Experimental	48	52	−3.4	18.7	-	-
	FE	45	59	−3.6	18.3	-	-
	Difference (%)	7	12	6	2	14	8
Berleburger 15 mm	Experimental	54	53	−2.5	20.1	-	-
	FE	50	53	−2.8	19.9	-	-
	Difference (%)	8	0	11	1	18	8
Holywell 55 mm	Experimental	51	50	−6.7	23.8	-	-
	FE	52	56	−6.6	23.1	-	-
	Difference (%)	2	11	2	3	14	7
Holywell 35 mm	Experimental	57	48	−5.8	25.1	-	-
	FE	59	51	−5.7	24.5	-	-
	Difference (%)	3	6	2	2	9	4
Holywell 15 mm	Experimental	63	45	−4.1	27.1	-	-
	FE	62	44	−4.3	25.6	-	-
	Difference (%)	2	2	5	6	6	4

5. Discussion

Microfoam models provided a good fit to the experimental shockpad data, particularly for the Berleburger shockpad. Although the Holywell shockpad exhibited more traditional hyper-elastic and viscoelastic behaviour, the microfoam model performed better overall than a more traditional hyper-elastic and viscoelastic Bergstrom-Boyce model. Further work is needed to understand the reasons for the differences in loading response for the two shockpads. Using the AAA test device to experimentally capture the stress-strain behaviour of the two shockpads under high strain rates was successful in helping to define these material models. Although the predictive capabilities of the two models have not been addressed, the ability of the models to fit not only to the impact drop height on which they were based, but also to lower energy impact drop heights, has been demonstrated. In reality, shockpads are exposed to a wide range of impact energies and a model that can estimate the response over this full range through a single set of measurements is relevant in optimising the design of 3G turf surface systems.

The AAA test can be very limited in terms of output if only considering the shock absorption and vertical deformation values as standalone measures of surface characteristics. Extracting the raw data from each of the drops allowed for a deeper analysis to be completed. Calculation of the corresponding stress and strain not only provided more information about the whole material response but also a way to use the data for material model calibration. This proved key in creating material models for both shockpads as using data recorded at slower strain rates, such as in the study by Mehvarar et al. [20], did not yield a material model that could successfully be used within AAA simulations.

The three regions of shockpad deformation, as described in the literature [13,14], were observed in the experimental stress-strain response at higher drop heights. An initial period of low stiffness and high deflection was followed by a transition region and finally an increase in stiffness at higher strains as the air was squeezed out of the system and rubber-on-rubber contact ensued. At lower drop heights this was less evident and the response was more linear in nature as the shockpad did not experience sufficient strain to enter the second or third stages of compressive behaviour. Interestingly, the loading response for the Berleburger shockpad showed undulations in the loading curve from the AAA impact that was not seen in either the unloading, or in the Holywell shockpad.

Repeatability between drops of the same height remained high as the shockpads remained in their elastic region even at the highest drop heights. Thus, a single drop could be used to represent the shockpad response. The Berleburger shockpad was stiffer and had higher peak accelerations associated with all impacts at different heights when compared to the Holywell shockpad. This may be due to the reduced thickness of the Berleburger shockpad, resulting in the mass and impact foot having a shorter deceleration distance. The strains for each shockpad remained similar despite the difference in thickness, suggesting that thickening of the Berleburger shockpad may lead to better shock absorbing properties.

Calibration of the material models to the experimental data was a multi-step process (Figure 7). The microfoam material model proved to be the more accurate as it allowed for the cellular characteristics of the shockpads to be captured. Relative density is an important characteristic for cellular materials as it expresses the density of the cellular material in relation to the dense solid [24]. The model was able to incorporate the relative density of the shockpads and thus make a distinction between the density of the solid compared to that of the cellular material. As this value could be estimated for both the shockpads, the variable could be fixed whilst optimizing for the remaining model coefficients.

The FE simulation was deemed accurate if the mean error across all comparative measures was less than 10%, with no individual error greater than 15% (Table 3). These values were chosen for a number of reasons but principally due to the experimental values used in the error assessment also being subject to errors. There was an uncertainty associated with the shockpad thickness values due to their construction method and granular nature (standard deviation from 15 thickness measurements was 0.1 mm for the Berleburger and 0.9 mm for the Holywell; Table 1). This has the potential to influence the experimental strain calculations and the outputs from the simulation. The AAA test uses an accelerometer embedded within the 20 kg drop mass to determine the shock absorption, vertical deformation and energy restitution, and in this study the accelerometer output was also used to determine the experimental stress–strain response of the shockpad. The main errors that result from use of the accelerometer are in numerical integration of the acceleration data to determine velocity and displacement, and in determining the instant and conditions at impact. Some measures of these uncertainties can be inferred from the FIFA Quality Concept [21]. In particular, these standards only require vertical deformation to be reported to the nearest 0.5 mm, corresponding to an uncertainty of 7–20% for the experimental values in this study (Table 3). In addition, the standards specify a range for the required impact velocity of 1.02 to 1.04 m/s (for the 55 mm drop height). A sensitivity analysis conducted on the FE simulation, where the impact velocity was increased over this range, found a 1% reduction in shock absorption (corresponding to a 2% error in the FE simulation value) and no detectable change in vertical deformation, energy restitution or contact time. The second main reason for the chosen error values relates to the levels of accuracy reported in a previous study that simulated the AAA drop test onto athletic track materials [15,16]. This previous study used a less extensive error analysis than performed here, however, comparable errors of up to 8% in shock absorption were reported and deemed acceptable for the simulated response.

Validation of the material model was completed by running FE simulations for different drop heights and comparing the results to the experimental data. The results showed a good match to the stress-strain response for the lower drop heights. This result is somewhat unsurprising given that the lower drop heights tended to follow the same loading—unloading curve as the higher impacts. The energy restitution was poorly estimated for the higher drop heights due to the rebound velocity being much higher in the FE simulation than that of the experimental data. This suggests more energy needs to be taken out of the system. Attempts to further adjust the material models to allow for more energy to be absorbed caused the maximum force to drop and the vertical deformation to increase beyond the 15% error margin. Therefore, optimisation of all parameters requires a compromise to be made between making the models stiff enough to stop excessive deformation, but also compliant enough to absorb the impact forces and dissipate the impact energy.

More accurate deformation responses for the shockpads under the lower drop heights could have been achieved through calibration of the models to the stress–strain data for these drop heights. Calibration to the 55 mm drop height allowed for all three phases of shockpad response to be captured.

Lower drop heights, particularly under 35 mm, were linear in nature and therefore were unlikely to be successful in estimating the shockpad response for higher drop heights. When considering the intended future application of this modelling work, i.e., to model the whole 3G turf surface system under human loading conditions, the stress at the shockpad level will be more variable than for the controlled conditions of this study. Therefore, having a model that can fit the shockpad response for a range of drop heights becomes much more desirable. While the 55 mm drop height (designed to emulate the impact of heel strike running on 3G turf) is likely to represent the maximum stress conditions the shockpad would be exposed to [26].

The FIFA test measurements of shock absorption, energy restitution and vertical deformation allow for easy evaluation of the models and comparison of the shockpad responses. However, they are reliant on only a few discrete values taken from the impact data. Using only these values it would be easy to miss any differences in the loading curves between the two shockpads. Therefore, creating a simulation capable of reproducing acceptable values only for the FIFA test criteria may not result in an accurate representation of the full shockpad response. Inclusion of strain RMSE as an assessment criterion was important in giving a measure of how well the simulation captured this full response.

6. Conclusions

The modelling of elastomeric materials is complex with limited material models at the users' disposal. This study has shown how data taken from a non–standard material test method (the AAA; [21]) can be used to successfully calibrate microfoam material models for two shockpads. The use of a foam model over a model for solid (rubber) materials performed better in describing the behaviour of both shockpads even though one shockpad demonstrated a more traditional hyper-elastic experimental response. Results from the AAA simulation demonstrated a good fit to the experimental data across most measurement variables and all drop heights. Future work should focus on improving the accuracy of the models as well as extending their predictive capacity, for example, to estimate the effects of altering shockpad thickness.

Author Contributions: All authors were involved in conceiving and designing the study; D.C. performed the experiments and analysed the data; All authors were involved in writing the paper.

Conflicts of Interest: The authors declare no conflict of interest.

References

1. Sport England Artificial Surfaces for Outdoor Sport: Updated Guidance for 2013. Available online: https://www.sportengland.org/facilities-planning/ (accessed on 28 January 2018).
2. Sport England. Sport Scotland Synthetic Turf Pitch User Survey. Available online: https://sportscotland.org.uk/ (accessed on 28 January 2018).
3. Fédération Internationale de Football Association FIFA Quality Programme for Football Turf. Available online: www.FIFA.com (accessed on 23 February 2018).
4. Gibson, O. The Guardian Online: FA Reveals Its 2020 Vision: Football Hubs and 3G Pitches for All. Available online: https://www.theguardian.com/football (accessed on 22 June 2016).
5. Dixon, S.; Fleming, P.; James, I.; Carre, M. *The Science and Engineering of Sports Surfaces*, 1st ed.; Routledge: Oxford, UK, 2015; ISBN 978-0-415-50092-0.
6. Fleming, P.; Ferrandino, M.; Forrester, S. Artificial Turf Field-A New Build Case Study. *Proc. Eng.* **2016**, *147*, 836–841. [CrossRef]
7. McLaren, N.; Fleming, P.; Forrester, S. Artificial grass: A conceptual model for degradation in performance. *Proced. Eng.* **2012**, *34*, 831–836. [CrossRef]
8. Smith, L.P. *The Language of Rubber: An Introduction to the Specification and Testing of Elastomers*; Butterworth-Heinemann: London, UK, 1993; ISBN 10: 0750614137.
9. Harper, C.A. Chapter 4: Elastomers. In *Handbook of Plastics Technologies*; McGraw-Hill: London, UK, 2006; ISBN 9780071460682.

10. Mills, N. *Polymer Foams Handbook Engineering and Biomechanics Applications and Design Guide*; Butterworth-Heinemann: London, UK, 2007; ISBN 978-0-7506-8069-1.

11. Kossa, A.; Berezvai, S. Visco-hyperelastic Characterization of Polymeric Foam Materials. *Mater. Today Proc.* **2016**, *3*, 1003–1008. [CrossRef]

12. Verdejo, R.; Mills, N.J. Heel-shoe interactions and the durability of EVA foam running-shoe midsoles. *J. Biomech.* **2004**, *37*, 1379–1386. [CrossRef] [PubMed]

13. Anderson, L. Elastomeric Shockpads for Outdoor Synthetic Sports Pitches, Ph.D. Thesis, Loughborough University, Loughborough, 2007.

14. Allgeuer, T.; Torres, E.; Bensason, S.; Chang, A.; Martin, J. Study of shockpads as energy absorption layer in artificial turf surfaces. *Sport. Technol.* **2008**, *1*, 29–33. [CrossRef]

15. Andena, L.; Briatico-Vangosa, F.; Ciancio, A.; Pavan, A. A finite element model for the prediction of Force Reduction of athletics tracks. In *Procedia Engineering*; Elsevier B.V.: Sheffield, UK, 2014; Volume 72, pp. 847–852.

16. Andena, L.; Briatico-Vangosa, F.; Cazzoni, E.; Ciancio, A.; Mariani, S.; Pavan, A. Modeling of shock absorption in athletics track surfaces. *Sport. Eng.* **2015**, *18*, 1–10. [CrossRef]

17. Farhang, B.; Araghi, F.R.; Bahmani, A.; Moztarzadeh, F.; Shafieian, M. Landing impact analysis of sport surfaces using three-dimensional finite element model. In *Proceedings of the Institution of Mechanical Engineers, Part P: Journal of Sports Engineering and Technology*; SAGE Publications: Thousand Oaks, CA, USA, 2015; pp. 1–6.

18. Thomson, R.D.; Birkbeck, A.E.; Lucas, T.D. Hyperelastic modelling of nonlinear running surfaces. *Sport. Eng.* **2001**, *4*, 215–224. [CrossRef]

19. Dal, H.; Kaliske, M. Bergström-Boyce model for nonlinear finite rubber viscoelasticity: Theoretical aspects and algorithmic treatment for the FE method. *Comput. Mech.* **2009**, *44*, 809–823. [CrossRef]

20. Mehravar, M.; Fleming, P.; Cole, D.; Forrester, S. *Mechanical Characterisation and Strain Rate Sensitivity of Rubber Shockpad in 3G Artificial turf*; Association for Computational Mechanics in Engineering: Cardiff, Wales, UK, 2016; pp. 8–11.

21. Fédération Internationale de Football Association FIFA Quality Concept for Football Turf: Handbook of Test Methods. Available online: www.FIFA.com (accessed on 23 February 2018).

22. Fédération Internationale de Football Association FIFA Quality Programme for Football Turf—Handbook of Requirements. Available online: https://football-technology.fifa.com/en/media-tiles/football-turf-handbook-of-requirements-2015/ (accessed on 23 February 2018).

23. BSW Berleburger Schaumstoffwerk GmbH Regupol®Elastic Shock Pads for Artificial Turf. Available online: https://www.berleburger.com/en/products/ (accessed on 10 July 2017).

24. Solorzano, E.; Rodriguez-Perez, M.A. Part IV Cellular Materials. In *Structural Materials and Processes in Transportation*; John Wiley & Sons: Hoboken, NJ, USA, 2013; pp. 371–374. ISBN 9783527649846.

25. Bijarimi, M.; Zulkafli, H.; Beg, M.D.H. Mechanical properties of industrial tyre rubber compounds. *J. Appl. Sci.* **2010**, *10*, 1345–1348. [CrossRef]

26. Wang, X. Advanced Measurement for Sports Surface System Behaviour under Mechanical and Player Loading, Ph.D. Thesis, Loughborough University, Loughborough, UK, 2013.

Article

Low-Velocity Impacts on a Polymeric Foam for the Passive Safety Improvement of Sports Fields: Meshless Approach and Experimental Validation

Francesco Penta [1], Giuseppe Amodeo [2,*], Antonio Gloria [3], Massimo Martorelli [1], Stephan Odenwald [2] and Antonio Lanzotti [1]

[1] Department of Industrial Engineering, Fraunhofer Joint Lab Ideas, University of Naples Federico II, 80125 Naples, Italy; francesco.penta@unina.it (F.P.); massimo.martorelli@unina.it (M.M.); antonio.lanzotti@unina.it (A.L.)

[2] Institute of Lightweight Structures, Technische Universität Chemnitz, 09001 Chemnitz, Germany; stephan.odenwald@mb.tu-chemnitz.de

[3] Institute of Polymers, Composites and Biomaterials, National Research Council of Italy, 80125 Naples, Italy; angloria@unina.it

* Correspondence: giuseppe.amodeo@s2015.tu-chemnitz.de; Tel.: +49-371-531-30363

Received: 31 January 2018; Accepted: 12 July 2018; Published: 18 July 2018

Abstract: Over the past few years, foam materials have been increasingly used in the passive safety of sport fields, to mitigate the risk of crash injury. Currently, the passive safety certification process of these materials represents an expensive and time-consuming task, since a considerable number of impact tests on material samples have to be carried out by an *ad hoc* testing apparatus. To overcome this difficulty and speed up the design process of new protective devices, a virtual model for the low-velocity impact behaviour of foam protective mats is needed. In this study a modelling approach based on the mesh-free Element Galerkin method was developed to investigate the impact behaviour of ethylene-vinyl acetate (EVA) foam protective mats. The main advantage of this novel technique is that the difficulties related to the computational mesh distortion and caused by the large deformation of the foam material are avoided and a good accuracy is achieved at a relatively low computational cost. The numerical model was validated statistically by comparing numerical and experimental acceleration data acquired during a series of impact events on EVA foam mats of various thicknesses. The findings of this study are useful for the design and improvement of foam protective devices and allow for optimizing sports fields' facilities by reducing head injury risk by a reliable computational method.

Keywords: sports safety; impact testing; foam protective mats; EFG method

---

1. Introduction

Among all sports injuries, head injuries represent the most severe risks to athletes' health. Nowadays, serious or fatal accidents are not uncommon on sports fields: in 2015, a 21-year-old footballer died in Argentina's fifth tier after his head collided with a concrete wall at the edge of the pitch [1]. A similar case in 2008 involved a Croatian football player who suffered a fatal head injury after colliding with a barrier positioned about three metres from the sideline [2].

The protection of athletes when they collide with barriers is a critical issue for the improvement of passive safety in sports fields. The lack of safety requirements in current sporting regulations is considered the main cause of injuries or fatalities. For example, current football requirements prevent the risk of impact for the athlete only by prescribing a minimum distance of dangerous equipment from the boundary lines of the playing pitch. They do not make mention of the physical behaviours of players during their performance and the correlated potential impact energy of the athlete in overstepping the boundary lines of the regular pitch [3,4]. Moreover, they do not prescribe specific actions (e.g., by covering sports facilities with high energy-absorption materials) in attenuating the potentials correlated to accidental impacts of the player's head against external sports equipment.

To ensure the players' safety, the introduction of materials with high-energy absorption properties in the surrounding area of the playing field is of primary importance. The good cushioning performance of the polymeric foams, beyond a unique combination of properties such as light weight, good ageing, chemical resistance, and inertness, has ensured the application of these materials in protective sporting equipment over the past few years [5–7].

According to the standards [8–11], the evaluation of the impact performance for a protective device requires the use of an instrumented drop test rig, where a hemispherical mass falling on the prototype to be analysed reproduces the event of a head crash (the most life-threatening situation). The head injury risk is then assessed through "Head Injury Models" (HIM), which are based on the impacting mass acceleration vs. time curve acquired during the impact event [12]. This testing procedure requires a dedicated falling weight impact apparatus and reveals various difficulties including the friction of the falling group, the high number of samples required, and a long set-up time.

To overcome these problems, a numerical model for the impact behaviour of the foam protective devices is needed. However, the analysis of impact events on polymeric foam padding is a very complex mechanical issue. The nonlinear behaviour of the foam material, having rate-dependent viscoelastic properties, and the presence of impulsive forces make an exact analytic solution unattainable. Furthermore, when the phenomenon is analysed by a standard FEM approach, high foam deformations can lead to numerical instabilities and low accuracy of the numerical predictions [13].

This paper concerns the design practice of new protective devices. The absorption behaviour of ethylene-vinyl acetate (EVA) foam mats at different impact energies was analysed by a numerical approach based on the mesh-free Element Galerkin method [14–16]. As a case study, the impact behaviour of plates under different impact velocity is considered. The mechanical properties needed to define the constitutive model of the foam have been obtained by standard compressive quasi-static tests on small samples.

The proposed modelling practice was validated statistically on the basis of data obtained from low-velocity impact tests. The validation study was conducted, comparing the output results of the simulations with the experimental ones. To verify the validity of the simulations, as comparison parameters the missile acceleration curves, the "Coefficient of determination" (R^2), and the "Root Mean Square Error" (*RMSE*) were used.

2. Materials and Methods

Low-velocity impact tests were performed on EVA foam mats through a low-velocity impact testing apparatus designed and built at the Fraunhofer Joint Lab IDEAS of the University of Naples Federico II. The experimental setup adopted in the present work is shown in Figure 1.

According to the ASTM F1292 standard [11], a hemispherical missile with 160 mm diameter and a mass of 4.6 kg, equipped with 500 g uniaxial accelerometer, was dropped from a height (through a guidance system) on the foam pad, which was fixed by an adhesive tape on a steel anvil. In each test the fall height of the missile was measured by a laser displacement sensor located close to the missile release mechanism. A second laser displacement sensor, positioned near the specimen surface and triggered by the missile position, was used to measure the missile velocity just before impacting on the foam pad.

The material on which tests were performed was an EVA closed cell foam with a density of 35 kg/m^3 in two different thickness of 50 mm and 70 mm. EVA foam is a closed cell foam made from ethylene-vinyl acetate and blended copolymers. It shows a high level of chemical cross-linking and, therefore, can be classified as a semi-rigid foam with a fine uniform cell structure. Because of those properties, the EVA foam is frequently used in applications where both high impact and vibration absorption, suitability for thermo-forming and thermo-moulding, are needed. Compared to other materials used for these applications, such as polyethylene foams, EVA foam is softer, more resilient, and has greater recovery characteristics after compression.

Figure 1. Drop test rig at Fraunhofer JL IDEAS at CESMA, Univ. of Naples Federico II.

In a recent study by Lanzotti et al. [4], football player movements in common gaming actions were analysed. Movements were considered potentially damaging when they occurred near the boundary lines of the regular pitch, since they lead players to overstep the boundary lines in an uncontrolled manner with a residual kinetic energy. The players' impact velocities against objects placed close to the playing field were also measured. The expected values of the impact velocities for object placed at distances of 1, 2 and 3 m away from the field boundary were equal to 4.5 m/s, 5.2 m/s, and 6.2 m/s, respectively. These results were adopted as reference values for the initial velocities of the impacting missile in the explorative tests carried out in the present study for realistically reproducing head impacts of increasing severity. Deeper analysis is, however, needed to definitively choose a set of initial missile velocities that take into account the difference between real impacts of athletes and those reproduced by a stiff standard test rig.

All the impact tests were carried out according to the practice reported in [10]. To evaluate the correct fall heights, h_m, corresponding to the three above reported velocities, a statistical model of friction between the linear guide carriage and the drop assembly during the vertical fall, was implemented and used. Furthermore, for each specimen the corresponding critical fall height (*CFH*), namely the maximum fall height from which a life-threatening head injury would not be expected to occur, was calculated. For this purpose, the iterative statistical procedure of [17] was employed. At a velocity of 6.2 m/s, the test was not performed on the EVA sample with a thickness of 50 mm because the corresponding fall height was higher than the *CFH* of this kind of specimen. The acceleration–time curve was acquired for each test. From this curve, the injury parameters maximum acceleration a_{max} and head injury criterion (*HIC*) were derived. In particular, an Online FAMOS routine was used to calculate the HIC values for any time sub-interval $[t_i, t_f]$ of the overall time span of the impact. This procedure gave back the HIC score as the maximum of all those calculated and the corresponding time interval $[t_i, t_f]$.

Compression Tests

The material behaviour was also analysed through a series of compressive tests with quasi-static loads. As far as the specimen preparation, test loading application, and data analysis go, the EN ISO 844:2014 and ASTM D 1621 standards were followed.

Compression tests were performed on cubic specimens having a size of $50 \times 50 \times 50 \ \text{mm}^3$. Testing loads were applied on specimens by an electro-mechanical universal testing machine (INSTRON 5566, Norwood, MA, USA) digitally controlled by BLUEHILL 2 software. More specifically, the tests were carried out under displacement control, with longitudinal strain rates $\dot{\varepsilon}$ equal to 0.1, 2.0 and 10 min^{-1}. To get data on foam hysteretic behaviour, specimens were compressed until a nominal strain of 80% was reached and then unloaded with the same strain rate amplitude of the loading phase. During each test, the value of the applied load and the specimen height reductions were acquired by the machine control software to obtain the foam engineering stress-strain curve.

Some compressive stress–strain curves are reported in Figure 2, which provides information on the foam mechanical performance. The observed mechanical behaviour is consistent with that of flexible foam. Specifically, an initial linear region was evident and suggested a stiff mechanical response at the onset. This region was followed by a portion of the curve with a lower stiffness. A final stiff region of the stress–strain curve was observed as it is usually reported for flexible foams. The central region of the curve did not show a plateau (a zero-slope region) and was characterized by a lower stiffness compared with the other two regions of the graph. A compressive modulus of elasticity (E) of 0.4136 MPa was calculated from the slope of the initial linear region $(\varepsilon < 0.02)$ of the stress–strain curve. Furthermore, the results of Figure 2 reveal a stiffening effect when the strain rate is increased, which represents a typical aspect of the viscoelastic behaviour of the material.

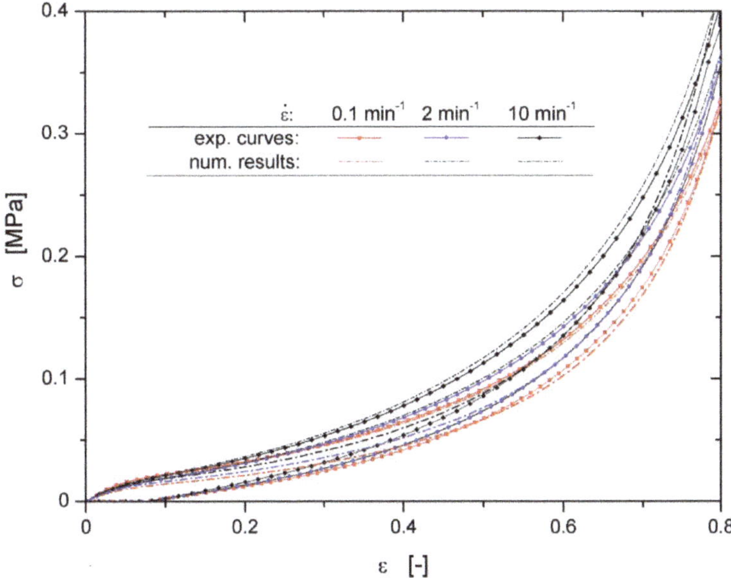

Figure 2. Loading—unloading compressive stress–strain curves of ethylene-vinyl acetate (EVA) specimens at different strain rates ($\dot{\varepsilon}_1 = 0.1 \ \text{min}^{-1}$, $\dot{\varepsilon}_1 = 2.0 \ \text{min}^{-1}$, $\dot{\varepsilon}_1 = 10 \ \text{min}^{-1}$).

3. Meshless Simulations

Several computational difficulties arose in the analysis of missile dynamic effects on the foam pad. The non-linear behaviour and the extremely large strain on the polymeric foam, together with the presence of an impulsive load, required a very small time step to be fixed to ensure the numerical solution of the pad equilibrium equations. For this reason, the central differencing scheme [18] was adopted to avoid the inversion of the pad stiffness matrix in each calculation step.

Moreover, the large deformations of the foam made the Lagrangian mesh extremely skewed and compressed. The remarkable distortion of the constitutive elements led to a reduction in the accuracy of the solution algorithm. To avoid accuracy reduction, periodic remeshing of the pad volume during the evolution of the simulation was needed. However, this solution produced a very high computational cost and the remeshing advantages were partially reduced from the degradation of accuracy due to the projection operation of the field variables from the old meshes to the new ones.

To overcome these difficulties, the Element-Free Galerkin (EFG) method of Belytchko et al. [16] was adopted to analyse the pad behaviour under missile impact. Using this method, the solution is constructed entirely in terms of a set of nodes and no elements are needed. The unknown displacement field $u^{(h)}$ is searched in the form given by

$$u^{(h)} = \boldsymbol{p}^T(x) * \boldsymbol{a}(x), \tag{1}$$

where $p(x)$ is the vector of the monomial basis functions and $a(x)$ is a vector of unknown parameters. These are determined at any point x by solving the following optimization problem:

$$minimize\ J = \sum\nolimits_{I=1} w(x - x_I) \left[\boldsymbol{p}^T(x) * \boldsymbol{a}(x) - u_I \right], \tag{2}$$

where u_I are the unknown values assumed by the displacement field at node I having coordinate x_I, $w(x - x_I)$ is a weighting function with compact support, and NP is the number of nodes within the support of w. When the solution to Equation (2) is substituted into Equation (1), we obtain for the approximation $u^{(h)}$ the expression

$$u_i^{(h)} = \sum\nolimits_{I=1} \varphi_I(x) u_I,$$

with $\varphi_I(x)$ being the shape functions of the EFG method given by

$$\varphi = [\varphi_1(x) \cdots \varphi_n(x)] = \boldsymbol{p}^T(x) A^{-1}(x) B(x),$$

where $A(x) = P^T W P$ and $B(x) = P^T W$, with

$$P = \begin{bmatrix} p(x_1) \\ \cdots \\ \cdots \\ p^T(x_n) \end{bmatrix} \text{ and } W = \begin{bmatrix} w(x - x_1) & 0 & \cdots & 0 \\ 0 & w(x - x_2) & \cdots & 0 \\ \cdots & \cdots & \cdots & \cdots \\ 0 & 0 & \cdots & w(x - x_n) \end{bmatrix}$$

The discretized equilibrium equations were derived by substituting the approximating function in the weak form of equilibrium conditions and carrying out numerical integration [15].

It is worth noting that the convergence of the Galerkin method chosen to discretize the problem at hand depended both on the approximation adopted for the unknown functions and the numerical integration of the weak form. To achieve the linear exactness solution in the Galerkin approximation, some integration constraints had to be met [19].

Therefore, the Lagrangian strain smoothing strategy described in [20] was used for the domain integration by Gauss quadrature. Finally, since the EFG shape functions are not interpolation functions, the essential boundary conditions were enforced by the transformation method proposed in [21].

Numerical Modelling

Drop impact tests on EVA square pads having thickness equal to 50 and 70 mm were simulated numerically using the commercial code LS-DYNA.

The calculation grids of the pads and the missile, as well as the background element mesh needed to carry on the Gaussian numerical quadrature, were built using the pre-processor HYPERMESH, starting from 3D models generated in the Solidworks geometrical modelling environment. In order to reduce the computational burden, the drop assembly geometry was simplified by considering only the hemispherical missile as an impacting object (with a mass equal to the fall group mass).

The hemispherical missile, made of aluminium alloy, was discretized using tetrahedral elements with an average size equal to 10 mm. Its deformations during the interaction with the pads were negligible compared to those of the pad. For this reason, the missile was modelled as a rigid body using the function MAT_RIGID (*MAT_020) [22].

To generate the nodes grid and the background mesh, the pads were divided into two distinct volumes and different node spacing was adopted for each volume. The first volume is the "impact zone," which has a prismatic shape with a base of size 200 mm × 200 mm, a height equal to the pad thickness, and includes the foam material more severely deformed by the missile actions (Figure 3). This volume was meshed with small hexahedral elements having edge size equal to 5 mm. The remaining portion of the pad was instead discretized with tetrahedral elements having a gradually increasing size from 5 mm at the inner boundary to around 10 mm at the external edges of the pad.

 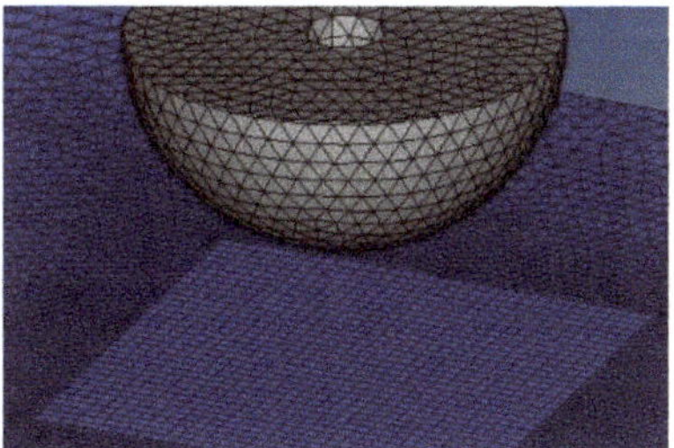

Figure 3. EFG model: Pad foam with the impact zone highlighted in green (**Left**) and missile (**Right**).

The MAT_LOW_DENSITY_FOAM (*MAT57) [22] was used as the foam constitutive model. This model is defined in terms of a foam compression curve, the Young's modulus, the Poisson ratio, and some constitutive parameters defining the hysteretic and viscoelastic aspects of the material response. The foam unloading behaviour was controlled by the hysteretic unloading factor, HU, and the SHAPE parameter, both reported in Table 1. HU ranged in value between zero and one. Low values of HU shifted the unloading path downward. The lowest value of HU did not sufficiently account for energy loss, so the SHAPE factor was adjusted to increase energy loss by further shifting the unloading curve down. Viscous effects were controlled by a DAMP factor. Increasing the DAMP values is equivalent to adding a material damper. The suggested range of DAMP was between 0.05 and 0.5.

Table 1. Material models used in the simulation and constitutive parameters. HU: hysteretic unloading factor; DAMP: viscous coefficient; SHAPE: shape factor for unloading.

Material Models	Density, ρ (kg/mm^3)	E (GPa)	ν	HU	DAMP	SHAPE
*MAT_RIGID	4.290×10^{-3}	6.9×10	0.33	/	/	/
*MAT_LOW_DENSITY_FOAM	3.500×10^{-5}	4.136×10^{-1}	0.33	0.01	0.35	1.00

The values of previous foam parameters were obtained by 'tuning' the results of a series of numerical simulations of the compressive tests with those obtained by the real compressive tests. A comparison among experimental results and numerical predictions carried out with optimal values of these constitutive parameters is given in Figure 2. Experimental data and numerical outcomes are in good agreement for the whole loading curves and for most of the unloading ones. The numerical model did not properly match the final part of the experimental unloading curves, since a part of the foam rearrangement takes place on a time scale greater than the compressive test time scale. However, in the following it will be shown that this does not influence the model accuracy in reproducing missile impacts on foam pads.

During the impact testing, penetration depths greater than the 80% of the pad thickness were observed. This clearly implies that for the higher impact velocities of the missile, the foam material is working in the densification phase of the stress–strain curve, where for small variations in the foam deformation the values of the compressive stress increase exponentially. However, by means of the compression tests, the real stress–strain curve of the foam was acquired only up to a nominal strain equal to 80%; after this threshold value of the strain specimen buckling started to occur.

Therefore, to carry out the numerical impact modelling with higher velocities, the Du Bois extrapolation to the densification region [23] was adopted for the EVA foam. Some results of a numerical simulation carried out by the developed model are reported in Figure 4, where the deformed shape of a 50 mm pad and the through-the-thickness distribution of the Von Mises stress are shown.

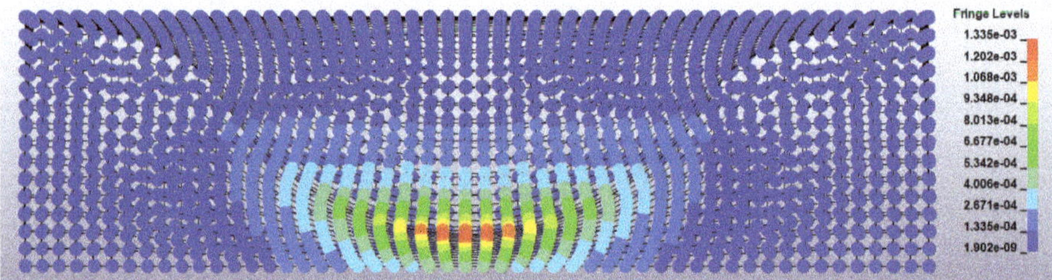

Figure 4. Mid-section deformed shape of the impact zone and through-the-thickness von Mises stress distribution (70 mm pad—units, GPa).

4. Results

In Figure 5, the simulated and experimental acceleration vs. time curves of the hemispherical missile impacting the EVA foam 50 mm at the velocity of 4.5 m/s are compared. In all the examined cases, to validate the numerical model using the statistical indexes $RMSE$ and R^2, only the most relevant part of the acceleration curves is used. More specifically, only the acceleration values belonging to the time range $[t_i, t_f]$ derived from the algorithm used to calculate the HIC index were considered. Actually, this is the most injurious portion of the impact waveform, where the acceleration reached the highest values.

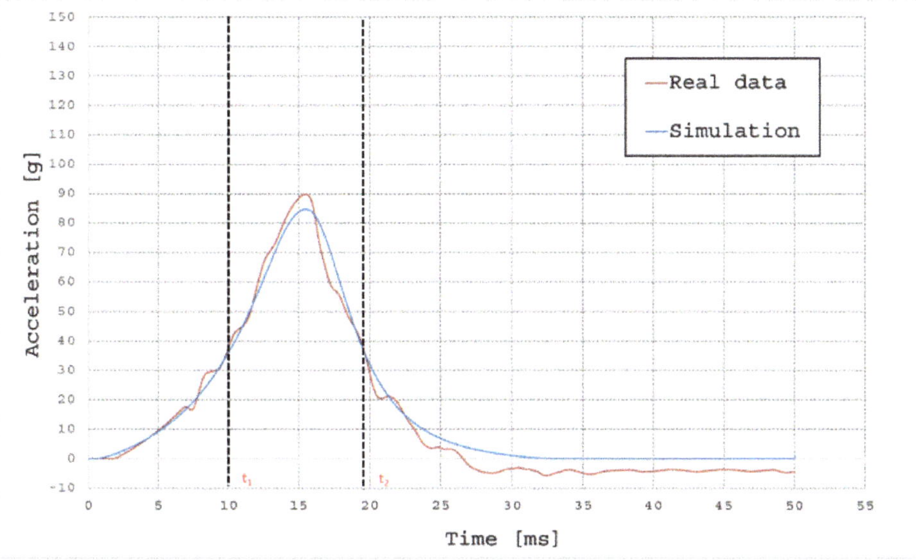

Figure 5. Comparison between the acceleration curve predicted by the simulation and the experimental curve for the EVA 50 mm at 4.5 m/s.

An overview of the simulated acceleration curves and the experimental ones for EVA samples with a thickness of 70 mm is shown in Figure 6. Instead, in Table 2, for each testing value of the impacting missile velocity V_{test} and for each foam plate thickness, the drop height H_m, the critical drop height CDH, the time interval $[t_i, t_f]$ where the comparison of the simulated and real test data is carried out, the maximum values of the acceleration reached in the real tests ($a_{max,r}$) and in the simulated ones ($a_{max,s}$), the HIC values, (HIC_r) and (HIC_s), respectively, evaluated by experimental and numerical data, and finally the values of the two statistical indices, $RMSE$ and R^2, adopted to evaluate the predictive accuracy of the model are listed. These indices were calculated using data sampled from the experimental and numerical acceleration curves with two frequency values, a choice derived from the necessity of also checking the veracity of the statistical validation. Actually, this latter condition is verified when the values of the two indexes calculated under different sampling frequencies are very close.

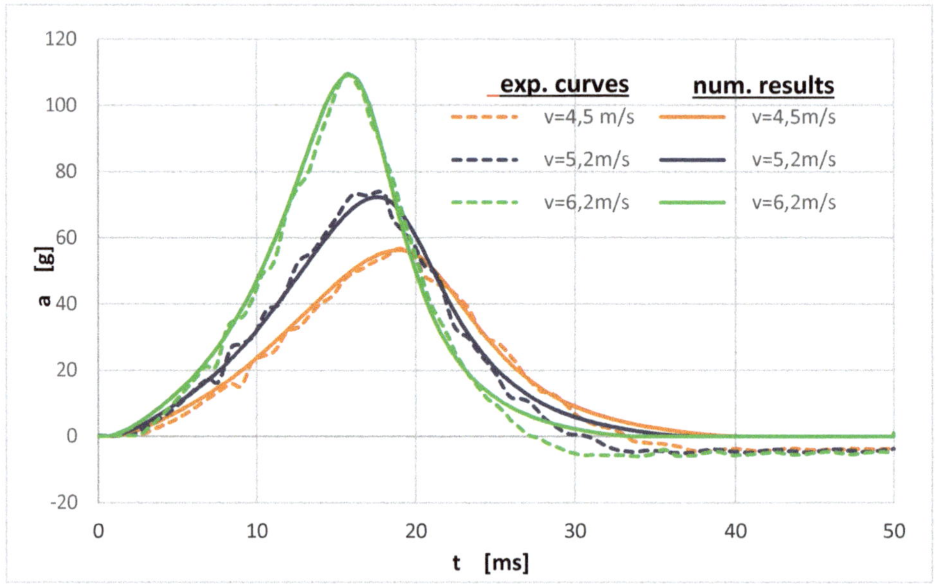

Figure 6. Experimental and simulated acceleration curves of the missile impacting the EVA 70 mm pad with initial velocities equal to 4.5 m/s, 5.2 m/s, and 6.2 m/s.

Table 2. Collected data from the impact test, the simulation outputs, and the statistical analysis. EVA: ethylene-vinyl acetate; HIC: head injury criterion.

Specimen Geometry	V_{test} (m/s)	H_m (m)	t_i (ms)	t_f (ms)	$a_{max,r}$ (g)	$a_{max,s}$ (g)	HIC_r	HIC_s	Sampling Time (ms)	RMSE	R^2
EVA foam 50 mm thickness (CFH = 1858 m)	4.5	1.302	10	19.5	94.3	83.8	312	297	0.25	4.50	0.93
									0.50	4.48	0.93
	5.2	1.709	10	16	163.0	98.1	599	375	0.25	23.68	0.24
									0.50	23.71	0.24
EVA foam 70 mm thickness (CFH = 3034 m)	4.5	1.302	10.5	25	59.2	55.2	173	174	0.25	1.53	0.97
									0.50	1.56	0.97
	5.2	1.709	10.5	23	78.6	72.6	299	292	0.25	2.86	0.95
									0.50	2.84	0.95
	6.2	2.387	10.5	20	118.0	107.9	589	594	0.25	2.63	0.98
									0.50	2.69	0.98

Further analysis was conducted for the EVA mat 50 mm at the impact test velocity of 5.2 m/s. In this case, the comparison between the simulated and real acceleration curves was carried out separately for the initial, central, and final regions of the diagram shown in Figure 7.

The results of the statistical analysis conducted in each of these regions are reported in Table 3. It is evident in this table that the accuracy of this analysis is very different from that of the previous ones. The values of the *RMSE* and R^2 indexes for the central region of the acceleration curve clearly indicate a poor correlation.

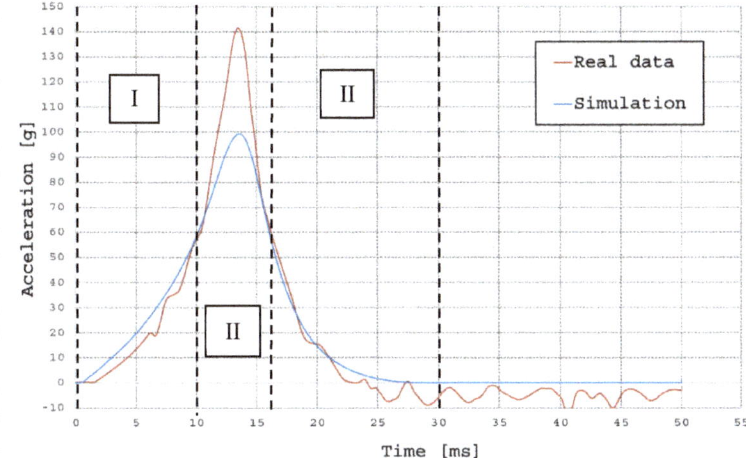

Figure 7. Comparison between the acceleration curve predicted by the simulation and the experimental curve for the EVA 50 mm at 5.2 m/s.

Table 3. Statistical analysis of the missile acceleration curve for the EVA 50 mm at 5.2 m/s.

Acceleration Time-History Region	t_i (ms)	t_f (ms)	Sampling Time (ms)	RMSE	R^2
Initial region (I)	0	10	0.25	4.35	0.93
			0.50	4.32	0.94
Central region (II)	10	16	0.25	23.68	0.24
			0.50	23.71	0.24
Final region (III)	16	30	0.25	4.32	0.94
			0.50	4.27	0.94

5. Discussion

As shown by the statistical analysis, the simulation succeeded in reproducing the physical phenomenon of a low-velocity impact test. The correct estimate of the a-max and HIC values by the simulation leads us to assess the risk of an athlete suffering an injury following a head impact, or in other words allows for ensuring an adequate level of protection for the athlete. The two curves of accelerations (real and simulated) show a similar trend, with discrepancies only in the last part of the impact event, where the simulation curve goes to zero and the real acceleration assumes negative values. This observed difference is due to the oscillations of the piezoelectric accelerometer before returning to zero values. This means that the simulations were performed correctly.

The values of the indexes *RMSE* and R^2 in Table 1 confirmed the good correlation between the experimental results and the simulated ones, except for the 50 mm EVA foam, which at an impact velocity of 5.2 m/s showed a remarkable difference in the region where the material reached the peak of acceleration (see Figure 7 and Table 3).

Regarding the data in Table 2, it is worth observing that the test fall height for the EVA 50 mm corresponding to the impact velocity of 5.2 m/s (equal to 1.709 m) appears to be very close to the CFH of the architecture (equal to 1858 m). Moreover, the simulation data show that the material reaches the densification stage during this test, while the behaviour of the material was modelled by the Du Bois model and not by the data obtained from the compression tests. Thus, the lack of knowledge about the real behaviour of the EVA foam in the densification zone is responsible for the poor correlation between the simulation and the test results observed in region II of Figure 7 and in Table 3.

The numerical simulations allowed us to investigate the behaviour of EVA polymeric foams subjected to a low-velocity impact. The main strength of the model is that its accuracy is insensitive to the large strains suffered by the foam material, which instead are responsible of numerical instabilities with a typical FEM analysis [14,21].

Furthermore, the developed numerical model may be employed to analyse the passive safety properties of shields for sports equipment made of generic polymeric foam and can be used when basic characteristic parameters of the materials under examination are known, such as the density, the Young's modulus, the Poisson ratio or the compressive stress–strain curve. The main application concerns the characterization of a foam mat under impact conditions. The use of the simulation in replacing the current testing standard is not feasible, as the deterioration of the absorption properties of the material in consecutive impacts is not contemplated in the model.

6. Conclusions

The purpose of this study was to create a versatile numerical model to simulate low-velocity impact tests on polymeric foam mats. The tool developed allows for the design and improvement of protective devices for sports safety without limitations due to the cost and time of an experimental test campaign.

A meshless method, namely the Element-Free Galerkin (EFG) method, has been adopted to overcome the limitations of the conventional finite element-based approaches and to achieve a high accuracy of the simulation. The main advantage of the proposed modelling procedure is that the foam constitutive properties needed for the analysis are just a few material parameters obtainable by standard laboratory testing.

A validation study has been carried out both by visual comparison of the impactor acceleration curves acquired during impact testing and numerical results and in terms of the statistical indexes R^2 and *RMSE*. The results of the study show that the adopted approach leads to very accurate results when the foam mechanical behaviour is not in the densification phase. When this happens, instead, numerical results are affected by errors related to the extrapolation of the foam compressive stress–strain curve. However, this is not a serious drawback since, to limit the accelerations due to impact to tolerable values, it is necessary to avoid the exponential increments of foam stiffness characteristic of its densification phase.

Author Contributions: Conceptualization, A.L., S.O.; Data curation, F.P., G.A., A.G. and M.M.; Funding acquisition, S.O. and A.L.; Investigation, F.P., G.A., A.G. and M.M. F.P. and G.A wrote the paper; F.P. performed numerical modelling; G.A. planned and carried out the low-velocity impact tests; A.G. planned and performed the compression tests; M.M. performed the optimisation of the geometric features and experimental setup; Writing—review & editing, F.P., A.G., G.A. and M.M.; Supervision, S.O. and A.L.

Funding: This work was supported by the State of Saxony, the European Union (European Social Fund—ESF), the University of Naples Federico II under a bilateral project and the European LLP (Lifelong Learning Program) under Erasmus Traineeship agreement.

Acknowledgments: This study has been carried out within the scientific cooperation between Chemnitz University of Technology (TUC) and University of Naples Federico II. The authors would like to thank the European LLP (Lifelong Learning Programme), which, through the project "Erasmus Traineeship," has funded the internship of the students Cecilia Maddaloni and Raffaele Moria at TU Chemnitz. Their work was essential for the collection and analysis of experimental data and the implementation of the simulation algorithm.

Conflicts of Interest: The authors declare no conflicts of interest related to this study.

References

1. Sheen, T. Emanuel Ortega Dead: 21-Year-Old Footballer in Argentina's Fifth Tier Dead after Suffering Horrific Head Injury after Collision with Concrete Wall. *Independent.* 14 May 2015. Available online: https://www.independent.co.uk/sport/football/news-and-comment/emanuel-ortega-dies-21-year-old-player-in-argentinas-fifth-tier-dead-after-suffering-horrific-head-10249706.html (accessed on 14 May 2015).
2. Ilic, I.; Oxley, S. Soccer-Croatian player dies after hitting head on wall in match. *Reuters.* 3 April 2008. Available online: https://uk.reuters.com/article/soccer-europe-croatia-death/soccer-croatian-player-dies-after-hitting-head-on-wall-in-match-idUKL0393548520080403 (accessed on 3 April 2008).
3. Odenwald, S.; Amodeo, G.; Costabile, G.; Lanzotti, A. Contribution to risk assessment in football by video analysis of overstepping boundary line events. *Sports Eng.* **2016**, *19*, 129–137. [CrossRef]
4. Lanzotti, A.; Costabile, G.; Annino, G.; Amodeo, G.; Odenwald, S. Video-Analysis of player's kinematics in running out of boundaries in Association football fields. *Procedia Eng.* **2016**, *147*, 234–239. [CrossRef]
5. Jenkins, M. *Materials in Sports Equipment;* Woodhead Publishing Limited: Cambridbe, UK, 2003.
6. Zhou, Y.J.; Lu, G.; Yang, J.L. Finite element study of energy absorption foams for headgear in football (soccer) games. *Mater. Des.* **2015**, *88*, 162–169. [CrossRef]
7. Coelho, R.M.; de Sousa, R.J.A.; Fernandes, F.A.O.; Teixeira-Dias, F. New composite liners for energy absorption purposes. *Mater. Des.* **2013**, *43*, 384–392. [CrossRef]
8. European Committee for Standardization. *Impact Attenuating Playground Surfacing, Determination of Critical Fall Height;* EN 1177:2008; CEN: Brussels, Belgium, 2008.
9. American Society for Testing and Materials. *Standard Test Method for Shock-Absorbing Properties of Playing Surface Systems and Materials;* ASTM F355-01; ASTM: West Conshohocken, PA, USA, 2001.
10. American Society for Testing and Materials. *Standard Specification for Impact Indoor Wall/Feature Padding;* ASTM F2440-04; ASTM: West Conshohocken, PA, USA, 2004.
11. American Society for Testing and Materials. *Standard Specification for Impact ttenuation of Surface Systems under and around Playground Equipment;* ASTM F1292-04; ASTM: West Conshohocken, PA, USA, 2004.
12. Cory, C.Z.; Jones, M.D.; James, D.S.; Leadbeatter, S.; Nokes, L.D.M. The potential and limitations of utilising head impact injury models to assess the likelihood of significant head injury in infants after a fall. *Forensic Sci. Int.* **2001**, *123*, 89–106. [CrossRef]
13. Auricchio, F.; Veiga, L.B.; Lovadina, C.; Reali, A. Stability of some finite element methods for finite elasticity problems, Mixed finite element technologies. In *CISM Courses and Lectures;* Springer: Wien, Austria, 2009; Volume 509, pp. 179–206.
14. Chen, Y.; Lee, J.; Eskandarian, A. *Meshless Methods in Solid Mechanics;* Springer Science & Business Media: New York, NY, USA, 2006.
15. Liu, G.R. *Meshfree Methods: Moving beyond the Finite Element Method;* Taylor & Francis: Boca Raton, FL, USA, 2009.
16. Belytschko, T.; Lu, Y.Y.; Gu, L. Element-Free Galerkin Methods. *Int. J. Numer. Methods Eng.* **1994**, *37*, 229–256. [CrossRef]
17. Schwanitz, S.; Costabile, G.; Amodeo, G.; Odenwald, S.; Lanzotti, A. Modelling head impact safety performance of polymer-based foam protective devices. *Procedia Eng.* **2014**, *72*, 581–586. [CrossRef]
18. Belytschko, T.; Liu, W.K.; Moran, B. *Nonlinear Finite Elements for Continua and Structures;* Wiley: Chichester, UK, 2000.
19. Chen, J.S.; Wu, C.T.; Yoon, S.; You, Y. A stabilized conforming nodal integration for Galerkin mesh-free methods. *Int. J. Numer. Methods Eng.* **2001**, *50*, 435–466. [CrossRef]
20. Chen, J.S.; Pan, C.; Roque, C.M.O.L.; Wang, H.P.A. Lagrangian reproducing kernel particle method for metal forming analysis. *Comput. Mech.* **1998**, *22*, 289–307. [CrossRef]
21. Li, S.; Liu, W.K. Meshfree and particle methods and their applications. *Appl. Mech. Rev.* **2002**, *55*, 1–34. [CrossRef]

22. Hallquist, J.O. *Keyword User's Manual Vol. II—Material Models*; LS-DYNA R8.0; Livermore Software Technology Corporation: Livermore, CA, USA, 2015.
23. Serifi, E.; Hirth, A.; Matthaei, S.; Mullerschon, H. Modelling of foams using MAT83–Preparation and evaluation of exfperimental data. In Proceedings of the 4th European LS-DYNA Users Conference, Ulm, Germany, 22–23 May 2003.

Article

Application of Auxetic Foam in Sports Helmets

Leon Foster [1,*], **Prashanth Peketi** [2], **Thomas Allen** [3], **Terry Senior** [1], **Olly Duncan** [4] and **Andrew Alderson** [4]

[1]   Centre for Sports Engineering Research, Sheffield Hallam University, Sheffield S1 1WB, UK; T.Senior@shu.ac.uk
[2]   Adidas Futures Team, Portland, OR 97217, USA; prashanth.peketi@adidas.com
[3]   School of Engineering, Manchester Metropolitan University, Manchester M1 5GD, UK; t.allen@mmu.ac.uk
[4]   Materials and Engineering Research Institute, Sheffield Hallam University, Sheffield S1 1WB, UK; Oliver.H.Duncan@student.shu.ac.uk (O.D.); a.alderson@shu.ac.uk (A.A.)
*   Correspondence: l.i.foster@shu.ac.uk; Tel.: +44-0114-225-3996

Received: 18 February 2018; Accepted: 19 February 2018; Published: 1 March 2018

Featured Application: Auxetic open cell foam was used as the comfort layer within a sports helmet and has been shown to reduce the severity of linear impacts. The work undertaken highlights the potential to further develop auxetic open cell foam as comfort layer and create more effective sporting protective equipment to reduce transmitted linear as well as rotational accelerations during an impact.

Abstract: This investigation explored the viability of using open cell polyurethane auxetic foams to augment the conformable layer in a sports helmet and improve its linear impact acceleration attenuation. Foam types were compared by examining the impact severity on an instrumented anthropomorphic headform within a helmet consisting of three layers: a rigid shell, a stiff closed cell foam, and an open cell foam as a conformable layer. Auxetic and conventional foams were interchanged to act as the helmet's conformable component. Attenuation of linear acceleration was examined by dropping the combined helmet and headform on the front and the side. The helmet with auxetic foam reduced peak linear accelerations ($p < 0.05$) relative to its conventional counterpart at the highest impact energy in both orientations. Gadd Severity Index reduced by 11% for frontal impacts (38.9 J) and 44% for side impacts (24.3 J). The conformable layer within a helmet can influence the overall impact attenuating properties. The helmet fitted with auxetic foam can attenuate impact severity more than when fitted with conventional foam, and warrants further investigation for its potential to reduce the risk of traumatic brain injuries in sport specific impacts.

Keywords: auxetic foam; helmet; concussion; sport; protection; impact attenuation

---

1. Introduction

In the United States alone, sport accounts for an estimated 3.8 million traumatic brain injuries (TBIs) annually [1]. TBIs, which include mild traumatic brain injuries (mTBIs) such as concussions, occur frequently and are severely underreported [2]. Symptoms include headaches and fatigue, which can lead to an increased risk of clinical depression, decreased satisfaction with life, and increased levels of disability [3,4]. Sahler and Greenwald [1] state that preventative measures against TBI most commonly take two forms: rule changes and innovations in impact protection in the form of helmets. Rowson et al. [5] examined two American football helmet designs over the course of 1.28 million impacts in games and practice, and found that helmet choice could reduce the number of concussions athletes endure. The helmet modulates energy transferred to the head, and the amount of energy transferred can differ by design. Helmet and head injury studies tend to explore the helmets' ability to reduce the risk of TBIs and mTBIs in sport [5–7].

Personal protective equipment (PPE), or other safety devices designed to prevent injuries arising from impact scenarios (short duration, high loading), aim to limit peak accelerations/forces, increase contact times, distribute loading, and perform energy attenuation. The performance of these safety devices is typically assessed by simulating an impact and measuring the ability to limit peak linear acceleration or force [8,9]. Helmet testing typically involves placing the helmet on the headform of an anthropomorphic test device (ATD) or a rigid headform, and using a drop-rig to transfer a known amount of impact energy. Accelerometers within the headform are used to record temporal accelerations and obtain impact characteristics including peak acceleration and impact duration [9,10]. Injury risk is typically quantified by assessing linear acceleration in combination with impact time, and relating to a severity index, such as the Head Impact Criterion (HIC) [11] and the Gadd Severity Index (GSI) [12]. GSI has been adopted in the protective equipment testing standards of the National Operating Committee on Standards for Athletic Equipment (NOCSAE) in the United States [10]. The GSI index is defined in Equation (1),

$$GSI = \int_{t_0}^{t_f} A^{2.5} dt \tag{1}$$

where A is the instantaneous resultant acceleration expressed as a multiple of gravity (g), dt is the time increment in seconds, and the integration bounds are from t_0 to t_f, the impact duration (determined by threshold value >0 g) [10,11]. New helmets are subjected to 16 impacts (44 to 60 J) onto a fixed rigid surface in six orientations. To pass the NOCSAE criterion, the GSI value for each impact must be below 1200, which signifies a non-injurious impact.

Helmets are a form of PPE used to protect the head and brain from injury. Impacts that contribute to head injury typically cause linear and rotational accelerations. The emphasis of attenuating rotational acceleration caused by oblique impacts has recently increased [13,14]. Oblique impacts increase the likelihood of brain injury [15] through mechanisms including: (i) rupturing arteries and bridging veins (causing subdural hematomas, SDH); and (ii) tearing neuro-connective tissue (causing diffuse axonal injuries, DAI) [16,17]. Mixed impacts (including rotational and linear accelerations) further increase the likelihood of fatal or coma inducing DAI and SDH [18–20]. Neither HIC nor GSI assess oblique or mixed impact testing, nor do they measure rotational acceleration, and historically helmet standards have been based on these indices [13,14]. Emphasis on linear (rather than rotational) acceleration has led to criticism of certifications and helmets alike [1].

To best attenuate linear acceleration, helmet materials are often viscoelastic, such as plastic based foams [21]. Compliant conformable foam is typically closest to the head (which is currently believed not to contribute to helmet function [22]), with stiffer, denser foam lining a hard-plastic shell [23]. A slip plane between a helmet's shell and deformable layer aims to reduce rotational acceleration and is now included in many cycling and snowsports helmets [24]. Evidence supporting slip plane technology's ability to reduce the likelihood of concussion is based on computational models, which have been criticized for not including physical validation [25–27]. A similar concept, increasing rotational deformation by reducing shear modulus, has reduced the severity of oblique impacts [22].

Despite helmet developments, there is evidence to suggest that the number of sport induced head injuries has not decreased. Firstly, Casson et al. [28] compared the number of concussed players in the National Football League (NFL) from the 1996–2001 and the 2002–2007 seasons. Suggested reasons for this lack of improvement in concussion rates are inadequate helmet design and certification. As recently as 2017, similar studies in snowsports saw helmet use rise dramatically after multiple high-profile deaths, serious injuries, and awareness campaigns, but with relatively little decrease in head injury rates [29,30].

Auxetic foams (with a negative Poisson's ratio) have shown higher energy absorption than their conventional counterparts [31–33], and can be tailored to test the effect of mechanical characteristics (i.e., Young's modulus and Poisson's ratio) in complex situations (i.e., impact tests) [34,35]. Lakes originally fabricated auxetic foams through a compression and heat treatment process [36]. Conventional open cell foams were compressed triaxially, to produce a re-entrant cellular structure, in molds with a volumetric compression ratio (VCR) between 1.4 amd 4 (VCR = initial volume/final volume). The molds were heated (163–171 °C) to soften the foam, and then cooled to room temperature to set the re-entrant cellular structure. Auxetic foams fabricated in this way typically have: (i) higher density due to volumetric compression; and (ii) quasi-linear compressive stress/strain relationship rather than cellular collapse beyond ~5% compression typical in conventional open cell foams [37].

Auxetic conversion methods have developed, including suggestions of multiple heating cycles [38], mold lubrication [39], cooling the foam slowly within the mold [40], and passing pins through the foam to control compression [41,42]. Pins allowed lateral stretching but through thickness compression, producing auxetic foams with comparable density and compressive stress strain relationships to their unconverted counterpart [34,43].

Open cell auxetic foams exhibit peak forces ~3 to ~8 times lower than their unconverted counterparts when impacted in scenarios similar to sporting standards [31,32,38,43,44]. Previous work has impact tested foams in isolation [33,44,45] or covered with thin polypropylene shells [31,32,38,43,44]. Comparisons between auxetic and conventional foams have not been made in more complex multi-material systems, such as helmets. Closed cell foams with low magnitude, incremental negative Poisson's ratio have been fabricated [46], but cells are reported to rupture during heat and compression. Therefore, the stability and practical benefits of closed cell auxetic foam is still in question and at present there are no studies that include impact testing of this type of foam (either as part of a complex system or in isolation).

This preliminary investigation aims to determine the effect on GSI when replacing an open cell conformable foam layer with an auxetic foam. Recent work assumes that the 'comfort layer' does not contribute to reduce impact severity [22]. Thermo-mechanically treating open cell foam alters its characteristics changing stress strain relationship, density, and Poisson's ratio. This work will explore if changes to the comfort layer can affect impact severity on a headform within a test helmet.

2. Materials and Methods

The effect of using auxetic foam within a typical sports safety helmet was investigated through fabrication, characterization, and comparison of conventional and auxetic foams within a helmet. The original helmet (Coolflo Batting Helmet, Rawlings, St. Louis, MO, USA) contained three layers: (1) an open cell polyurethane foam as a conformable layer acting as an interface between the head and the helmet; (2) a closed cell ethylene-vinyl acetate (EVA) foam; and (3) a semi-rigid plastic exterior. This helmet was chosen for its simple construction which allowed easy substitution of candidate foams to act as the conformable layer.

The exact grade of open cell polyurethane foam used as the native conformable foam layer in the helmet was unknown and it was not possible to source additional foam of this grade for auxetic conversions. The conformable layer was therefore replaced with an alternative foam (PUR30FR, Custom Foams) which is well established for auxetic research, to ensure consistent conversion in line with previous studies [34,38,44]. A sample of the native conformable foam and the un-converted PU R30FR foam (41 × 127 × 25 mm) was compression tested through thickness (25 mm dimension) with a uni-axial test device (Instron 3367, Norwood, MA, USA, with a 500 N load cell) to 50% compression (0.0083 s^{-1}). A linear trend line was fitted to the initial linear region of the stress vs. strain data to obtain Young's modulus (34.3 kPa, plateau at 3–3.5 kPa, ~10% strain) and check it was comparable to PU R30FR (~30 kPa, plateau at 3–3.5 kPa, ~5 to 10% strain).

The auxetic foam was fabricated using a thermo-mechanical method for uniform sheets (~350 × 350 × 20 mm) with compression controlled by pins in the initial heating cycle, previously developed, and described in detail [34,42]. The same foam (PUR30FR, Custom Foams), mold compression regimes (isotropic Linear Compression Ratio, LCR of 0.7, defined as the ratio of the compressed to initial dimension), and heat cycles (25 min then 15 min at 180 °C) were employed. The annealing cycle was reduced to 10 min at 100 °C. Two foam sheets were produced, the first used for establishing the setup and pilot testing, and the second for final testing, to limit degradation.

Prior to impact testing within the helmet, the converted foam was characterized in order to ensure auxetic characteristics similar to previously created samples [34]. A 200 × 20 × 20 mm auxetic sample cut from the center of the pilot sheet was subjected to a quasi-static tensile test (Instron 3369, Norwood, MA, USA, 10 kN load cell) up to 50% extension ($0.0017\,\text{s}^{-1}$). The sample was held in place with bonded card-board end tabs (Araldite, Super Strength). Four white pins were placed in a square (10 mm sides) on the front face of the sample, and their movement during testing was recorded using an HD camcorder (JVC Everio, 1920 × 1080 pixels). Lateral vs. axial strain data (Marker Tracking, MATLAB, Mathworks 2015b [44]) were fitted with linear trend lines (between 0 and 50% extension) to obtain Poisson's ratio.

Images of the foams and helmet can be seen in Figure 1 and the foam layer makeup is shown in Figure 2. Test foams were cut from the sheets in specific shapes using a bandsaw (Bauer Machinenbau). The forehead (Figure 1b) and side (Figure 1a,c) conformable foams were replaced with test foams. The top foam was only ~5 mm thick, and assumed to contribute less than other areas, to the helmet's ability to reduce impact induced acceleration. Therefore, the native foam at the top location was left in situ. Velcro strip tape was used to hold foam samples in place, for ease of attachment and removal. Locating marks were placed on the ATD headform to ensure consistency of helmet placement for each test. The properties of both helmets are tabulated in Table 1, along with the specification of the foams used in testing. Note that there is a larger area and hence volume of foam used in the side inserts when compared to the front inserts.

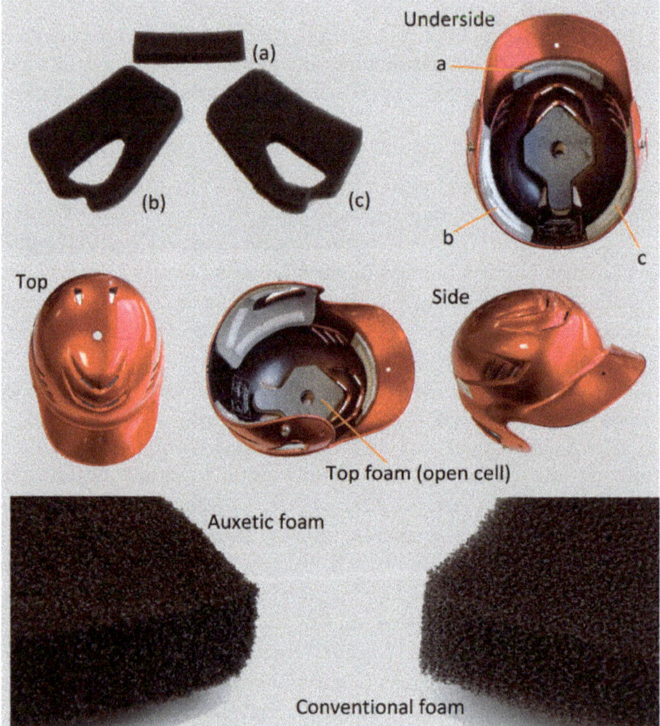

Figure 1. Foam inserts: (**a**) forehead, (**b**) right side, and (**c**) left side and views of the helmet showing native closed cell EVA foam and the top open cell foam in situ along with markers showing where inserts are placed. Also showing the foam structures for the auxetic (**bottom left**) and conventional (**bottom right**).

Table 1. Helmet and foam properties.

Location within Helmet	Foam Thickness (mm)		Replacement PU Foam Thickness (Conventional and Auxetic)	Mass (kg)		Conformable Foam Area of Largest Face (cm^2)	
	Native EVA	Native PU					
Top	11	5	n/a	Headform:	4.42	Front:	~63
Side	4	23	20	Helmet and foam:	0.54	Side:	~150
Forehead	6	23	20	Cradle:	0.1	-	-

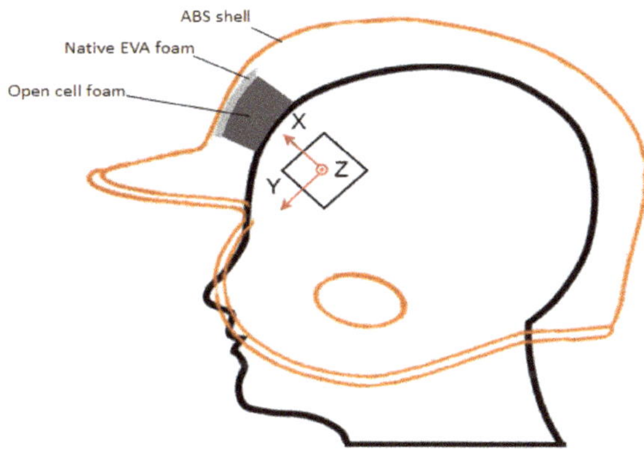

Figure 2. Schematic of the headform within the helmet showing position of accelerometer within headform and foam layers (z-axis coming out of the page).

An Anthropomorphic Test Device (ATD) headform (Hybrid III 50th Percentile Male, Humanetics) provided a surrogate head inside the helmet. To measure acceleration, a ±200 g triaxial accelerometer, fixed to an evaluation board (ADXL377, Analog Devices) was mounted within the headform, Figure 2. The accelerometer was wired to an NI USB-6210 Data Acquisition Device (National Instruments, Austin, TX, USA), sampling at 41,666 Hz. The accelerometer setup was calibrated using the protocols in the Federal Motor Vehicle Safety Standard 572 No. 208-E [47].

Testing was undertaken using a drop rig (Figure 3) with the helmet/headform in two orientations. Impact testing at multiple locations is typical of helmet test standards [10] and in this case was used to examine the severity of impact from a frontal and a side impact. The frontal and side impact tests engaged the forehead and side foams during the impact respectively.

Figure 3 shows both the frontal and side impact orientations, with the helmet landing on two thin (3 mm) polyurethane sheets, as used in the NOCSAE 001-13m15 standard [10]. These polyurethane sheets are used to prevent surface damage to the helmet. A range of drop heights (17, 25, 30, 37.6, 43, 50, 60, 70, and 80 cm) were tested, spanning impact energies of 8.3 to 38.9 J. These impact energies are below the 45 J maximum specified in the NOCSAE 022-10m12 [48] standard for the performance of baseball helmets (test helmet conforms to NOCSAE baseball standard). Impact energies were calculated based on the potential energy of the helmet/headform (mass ~5 kg). Five drop tests were completed at each drop height, for each foam type. The Gadd Severity Index was calculated for each impact. For frontal impacts, drop heights corresponding to impact energies of 8.3, 12.2, 14.6, 18.3, 20.9, 24.3, 29.2, 34.1, and 38.9 J were used. Side impacts were carried out at 12.2, 18.3, 20.9, and 24.3 J. The measuring range of the accelerometer in the z-axis prevented higher energy testing for the side impacts.

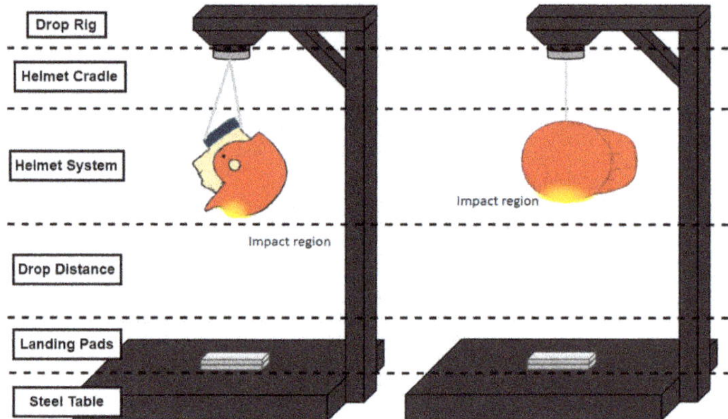

Figure 3. Schematic of the experimental setup depicting cradle to suspend the headform for frontal impact (**left**) and side impact (**right**).

Steps were taken to ensure consistency of helmet positioning on the ATD, orientation of the magnetic drop cradle, location of landing pads, and drop height. String was wrapped through an existing hole in the helmet, and under the chin of the headform, to secure it to the headform and ensure that the helmet did not slide loose from the headform mid-trial. To limit any effect of foam degradation, each sample was removed from the helmet and allowed to recover for 10 min before the subsequent drop. Resting time was calculated based on pilot testing with samples from the first converted foam sheet. Drops over a range of energies (14.6–29.2 J) were completed with increasing rest times for the auxetic and conventional foam helmets, until five sequential readings of similar 'peak' accelerations (±5%) were measured. Tests were carried at room temperature, and environmental effects such as humidity and temperature were limited by alternately testing each foam in the helmet (after five repeats at each drop height) during the same period.

Raw acceleration data was logged for each test and post-processed. To minimize any systematic noise within the signal a fast Fourier transform (FFT) was applied to the raw acceleration data. Prominent noise frequencies were then removed using a notch filter. The impact duration of a drop test was distinguished by the period of time the acceleration rose above a baseline of 1.5 g. The Gadd Severity Index for each impact was then calculated using Equation (1) [12].

3. Results

The fabricated auxetic foam was found to have a marginally negative tensile Poisson's ratio of −0.02, a Young's Modulus of 34 kPa, and density of 83 kg/m^3 which is comparable to previous work [34,38,44].

Resultant acceleration traces for frontal impacts with the conventional and auxetic foam inserts at low (12.3 J), medium (24.5 J), and high (38.9 J) energies are shown in Figure 4a–c. As expected, peak acceleration experienced by the headform increased with impact energy, and impact duration decreased. Figure 4d,e shows the resultant acceleration trace for side impacts. Conventional and auxetic foams perform in a similar fashion at low impact energies, but their performance appears to deviate at higher impact energy (29.2, 34.1, and 38.9 J). During a frontal impact at 38.9 J (Figure 4c) and side impact at 24.3 J (Figure 4e) auxetic foam inserts reduced the peak acceleration experienced by the headform. The mean Gadd Severity Index and standard deviation for each impact energy and foam inserts type has been summarized in Figure 5. Reflecting the acceleration traces in Figure 4, Gadd Severity Index increases with impact energy.

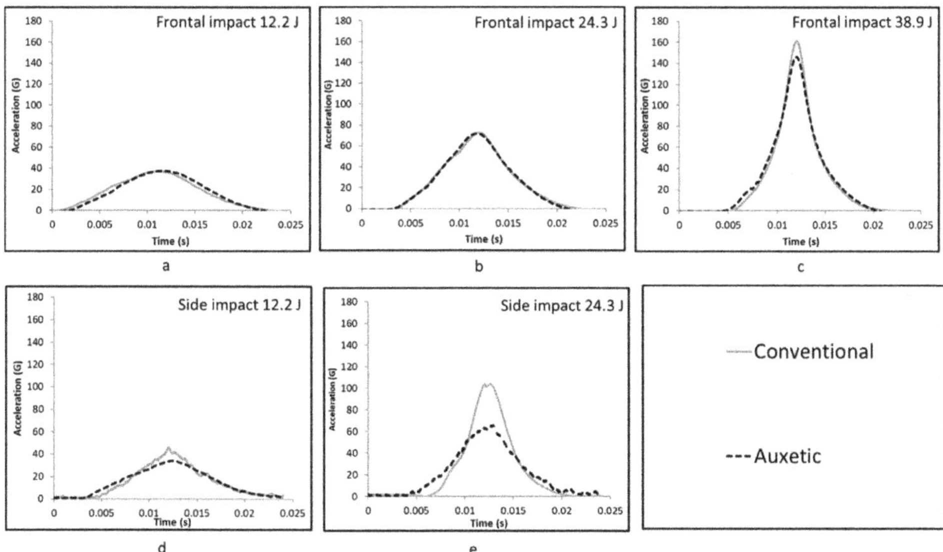

Figure 4. Acceleration traces for the conventional and auxetic foam inserts in the frontal impacts at (**a**) low (12.2 J), (**b**) medium (24.3 J), and (**c**) high (38.9 J) energies and for side impacts at (**d**) low (12.2 J) and (**e**) medium (24.3 J) energies.

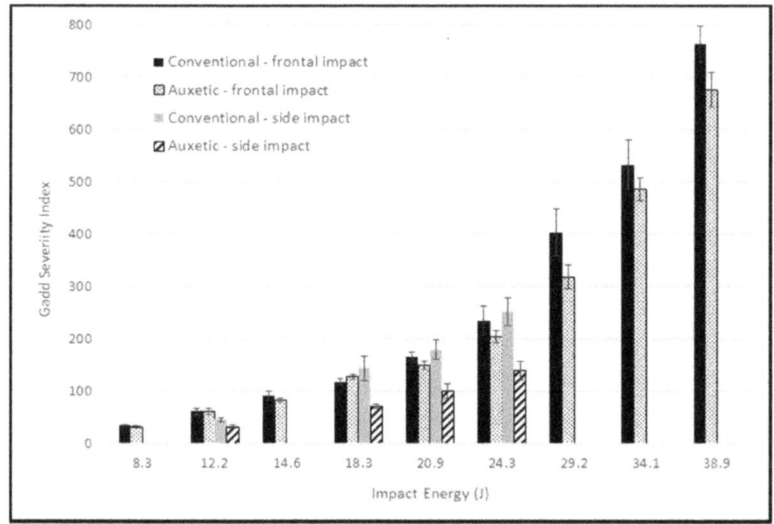

Figure 5. Helmet drop test severity for the different impact energies, showing both the frontal impact and side impacts (error bars ±1 s.d.).

The difference in Gadd Severity Index between conventional foam inserts and auxetic foam inserts for each impact energy is shown in Figure 6. The greater the positive value, the more the auxetic foam inserts reduces the GSI and a negative value indicates that the conventional foam inserts perform better. The general trend indicates the difference between foam types increases with impact energy.

Note that the open markers in Figure 6 show where the difference was not significant based on an unpaired *t*-test ($p > 0.05$).

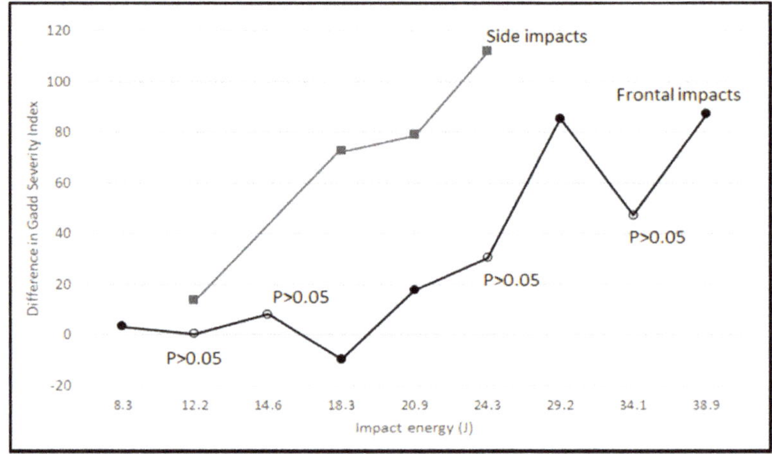

Figure 6. The difference in Gadd Severity Index between conventional foam inserts and auxetic foam inserts for the front and side impacts, labeled where these differences were not significant based on *t*-test $p > 0.05$ (open markers).

A factorial analysis of variance (FANOVA) was carried out for the data collected in (1) the frontal drop test and (2) the side drop test. In both cases, it was found that there was a significant effect for the impact energy, foam insert type, and also an interaction effect ($p < 0.05$ in all cases). This *p*-value of less than 0.05 indicates that the performance of the different foam insert varies depending on impact energy, and referring back to Figure 6, it appears that this interaction effect means that at lower impact energies, the performance of each foam type is similar but at higher impact energies they differentiate.

4. Discussion

A helmet with auxetic foam inserts significantly ($p < 0.05$) reduced the acceleration recorded by the headform accelerometer at the highest impact energy in both frontal and side impacts. In the frontal drop test at an impact energy of 38.9 J the auxetic foam reduced the GSI by 11.4%. In the side drop test at an impact energy of 24.3 J the difference in GSI was 44.2% between the foam types. At lower impact energies it appears that the difference between foam types is less, and only when the impact energy was above 18.3 J for the frontal drops did the auxetic foam reduce the GSI in comparison to conventional foam. For side drops at all impact energies tested, the auxetic foam performed better, reducing all GSI values. For the frontal drop test at the second highest impact energy (34.1 J), the auxetic foam inserts did seem to perform better than the conventional counterpart but this difference was not deemed significant at the 95% confidence level. The reason for this difference not being significant at 34.1 J is likely due to measurements errors, although there could be an underlying interaction effect of the foams, helmet, and headform at this impact energy which renders the auxetic foam less effective. In-depth further investigation, with more samples and tests scenarios, is required to fully understand the effectiveness of auxetic foams in comparison to their conventional counterpart across a range of impact energies.

In general, the findings are in line with previous work where auxetic foam tested in isolation or with thin polypropylene sheets attenuated linear impact acceleration to a greater extent at higher impact energies [34,38,44]. Allen et al. [38] stated that this behavior could be attributed to the auxetic foam's resistance to "bottoming out" during impact. 'Bottoming out' is the point at which an impact attenuating system becomes saturated due to high loading and the effectiveness of the system diminishes significantly. In the case of Allen et al. [38], this would be when the foam compresses to such an extent that it was no longer effective, and the impactor is essentially striking the anvil. At this point, the acceleration trace will show a 'spike' (short duration high peak).

In this study, any 'bottoming out' of the helmet system is hard to identify and assess due to the multi-layered foam construction of the helmet. The 'bottoming out' of the conventional open cell foam is likely to occur when the performance of the auxetic and conventional foam diverges. For the frontal impacts, deviation in impact severities occurs at an impact energy around 20.9 J, but for the side impacts this occurs at the lowest energy tested at 12.2 J. The reason why a conventional foam 'bottoms out' when compared to an auxetic foam is still unknown but could be due to the auxetic foams' deformation process and densification of the material occurring under the impact location [38].

Figure 6 illustrates that during the side impact tests the auxetic foam system appeared to produce significantly lower severity index over the range of energies tested. This significant improvement in the auxetic foam during the side drop test could be explained by the variation in the amount of foam within the helmet. The side conformable foam insert had approximately twice the amount of foam when compared to the front insert (Table 1). Therefore, during the side drop tests the increased amount of foam in the side insert appeared to exaggerate the energy attenuation ability of the auxetic foam. The conventional foam in the side impact test performed in a similar manner to the frontal drop test with similar GSI values. It appears that as the surface area and hence volume of the foam increases this also increases the ability of the auxetic foam to reduce GSI. However, the relationship between the auxetic foam's area/volume and its energy attenuating ability is unclear. The exact mechanism for why a larger amount of auxetic foam enhances its performance warrants further investigation but could be related to either the density of the auxetic foam, the curvature of the impacting objects, or the ability of the auxetic foam to densify under impact more effectively in a larger volume.

The thickness of the closed cell layer of native foam also varied within the helmet and was thicker in the front when compared to the side (~6 mm and ~4 mm respectively, Table 1). This means the contribution of the conformable layer during an impact may have been reduced at the front location. In general, the auxetic foam appeared to outperform or be equivalent to the conventional foam for both impact orientations.

Previous work [34,38,44] found much larger acceleration attenuation differences between the auxetic and conventional foams, on the scale of 300–800% compared to a maximum of 44% in this study. However, this reduction in the effectiveness of auxetic foams may be expected as they are no longer being tested in isolation, but as part of a multi-layered helmet. Findings agree with previous work [33,36], and it is notable that the conformable layer can significantly affect the helmet's performance. As the comfort layer can influence impact properties, the use of tuned auxetic foams could be used to enhance a helmet's performance. However, the effect of modifying the open cell layer within a helmet on the additional functions (i.e., fit and comfort) are yet to be assessed.

Anisotropic foams with low shear modulus could be used in helmets to specifically reduce rotational accelerations. Anisotropy and uniaxial Poisson's ratios can be increased in open cell foams by stretching parallel to cell rise during fabrication [34,35,49] rather than compressing (which typically reduces anisotropy and Poisson's ratio) [36]. Given that the open cell layer can reduce impact severity. Further work should expand upon reduced rotational accelerations exhibited by anisotropic foams with a low shear modulus and could be a way of reducing rotational acceleration in certain impacts, either as an alternative to or in parallel with slip plane technology [22,24]. One emerging area requiring further investigation is the use of 3D printed foams/structures which could enhance and tailor the impact performance of helmets further still [50]. This study is limited to one helmet design at two impacting sites using one grade of conventional and auxetic foam. Further work is required to replicate this experiment for a range of different helmets, foam grades, and impact scenarios across a number of sports. It would also be necessary to investigate how the tailoring of helmet liners influences ergonomic considerations, including pressure distribution, comfort, and fit.

5. Conclusions

This investigation revealed important advantages of the helmet containing auxetic foam to its conventional counterpart. The auxetic foam helmet attenuated linear acceleration significantly more than the conventional foam helmet for frontal impacts at the highest impact energy tested and for side impacts at all energies tested. The conventional foam in side impact tests appeared to perform in a similar manner to frontal drop tests at all energies, but auxetic foam in side impact tests produced significantly lower severity indices. Ultimately, this finding supports previous research indicating that auxetic foams could be utilized in sports safety applications and could lead into an interesting study to assess whether tailoring the open cell conformable layer in a traditional helmet by varying Poisson's ratio could reduce the risk of sport induced TBIs.

Author Contributions: Leon Foster, Thomas Allen, Andy Alderson, Terry Senior, and Prashanth Peketi conceived and designed the experiments, with Terry Senior designing and manufacturing all components required for drop testing. Prashanth Peketi carried out the experiment, wrote up the results, and analyzed the data. Leon Foster, Olly Duncan, and Prashanth Peketi wrote the paper which was edited by Thomas Allen and Andrew Alderson.

Conflicts of Interest: The authors declare no conflict of interest.

References

1. Sahler, C.S.; Greenwald, B.D. Traumatic brain injury in sports: A review. *Rehabil. Res. Pract.* **2012**, *2012*, 659652. [CrossRef] [PubMed]
2. McCrea, M.; Hammeke, T.; Olsen, G.; Leo, P.; Guskiewicz, K. Unreported concussion in high school football players: Implications for prevention. *Clin. J. Sport Med.* **2004**, *14*, 13–17. [CrossRef] [PubMed]
3. Stålnacke, B.; Björnstig, U.; Karlsson, K.; Sojka, P. One-year follow-up of patients with mild traumatic brain injury: Post-concussion symptoms, disabilities and life satisfaction at follow-up in relation to serum levels of S-100B and neuron-specific enolase in acute phase. *J. Rehabil. Med.* **2005**, *37*, 300–305. [CrossRef] [PubMed]
4. Guskiewicz, K.M.; Marshall, S.W.; Bailes, J.; McCrea, M.; Harding, H.P., Jr.; Matthews, A.; Mihalik, J.R.; Cantu, R.C. Recurrent concussion and risk of depression in retired professional football players. *Med. Sci. Sports Exerc.* **2007**, *39*, 903–909. [CrossRef] [PubMed]
5. Rowson, S.; Duma, S.M.; Greenwald, R.M.; Beckwith, J.G.; Chu, J.J.; Guskiewicz, K.M.; Mihalik, J.P.; Crisco, J.J.; Wilcox, B.J.; McAllister, T.W.; et al. Can helmet design reduce the risk of concussion in football? Technical note. *J. Neurosurg.* **2014**, *120*, 919–922. [CrossRef] [PubMed]
6. Kis, M.; Saunders, F.; ten Hove, M.; Leslie, J. Rotational acceleration measurements-evaluating helmet protection. *Can. J. Neurol. Sci.* **2004**, *31*, 499–503. [CrossRef] [PubMed]
7. Post, A.; Hoshizaki, T.B. Rotational acceleration, brain tissue strain, and the relationship to concussion. *J. Biomech. Eng.* **2015**, *137*, 030801. [CrossRef] [PubMed]
8. Rowson, S.; Duma, S.M. Development of the STAR evaluation system for football helmets: Integrating player head impact exposure and risk of concussion. *Ann. Biomed. Eng.* **2011**, *39*, 2130–2140. [CrossRef] [PubMed]
9. McIntosh, A.S. Biomechanical considerations in the design of equipment to prevent sports injury. *Proc. Inst. Mech. Eng. Part P* **2012**, *226*, 193–199. [CrossRef]
10. NOCSAE DOC. *001-13m15, Standard Test Method and Equipment Used in Evaluating the Performance Characteristics of Protective Headgear/Equipment*; National Operating Committee on Standards for Athletic Equipment: Overland Park, KS, USA, 2011.
11. Marjoux, D.; Baumgartner, D.; Deck, C.; Willinger, R. Head injury prediction capability of the HIC, HIP, SIMon and ULP criteria. *Accid. Anal. Prev.* **2008**, *40*, 1135–1148. [CrossRef] [PubMed]
12. Gadd, C.W. *Use of a Weighted-Impulse Criterion for Estimating Injury Hazard*; SAE Technical Paper No. 660793; SAE International: Warrendale, PA, USA, 1966.
13. King, A.I.; Yang, K.H.; Zhang, L.; Hardy, W.; Viano, D. Is head injury caused by linear or angular acceleration? In Proceedings of the IRCOBI Conference, Lisbon, Portugal, 25 September 2003.

14. Rowson, S.; Duma, S.M. Brain injury prediction: Assessing the combined probability of concussion using linear and rotational head acceleration. *Ann. Biomed. Eng.* **2013**, *41*, 873–882. [CrossRef] [PubMed]

15. Aare, M.; Kleiven, S.; Halldin, P. Injury tolerances for oblique impact helmet testing. *Int. J. Crashworth.* **2004**, *9*, 15–23. [CrossRef]

16. Gennarelli, T.A. Head Injury in Man and Experimental Animals: Clinical Aspects. *Acta Neurochir. Suppl.* **1983**, *32*, 1–13. [CrossRef] [PubMed]

17. Margulies, S.S.; Thibault, L.E. A proposed Tolerance Criterion for Diffuse Axonal Injuries in Man. *J. Biomech.* **1992**, *25*, 917–923. [CrossRef]

18. Dimasi, F.P.; Eppinger, R.H.; Bandak, F.A. Computational Analysis of Head Impact Response under Car Crash Loadings. In Proceedings of the 39th STAPP Car Crash Conference, San Diego, CA, USA, 8–10 November 1995; SAE 952718. SAE International: Warrendale, PA, USA, 1995. [CrossRef]

19. Ueno, K.; Melvin, J. Finite Element Model Study of Head Impact Based on Hybrid III Head Acceleration: The Effects of Rotational and Translational Acceleration. *J. Biomech. Eng.* **1995**, *117*, 319–328. [CrossRef] [PubMed]

20. Gennarelli, T.A.; Thibault, L.E.; Adams, J.H. Diffuse Axonal Injury and Traumatic Coma in the Primate. *Ann. Neurol.* **1982**, *12*, 564–574. [CrossRef] [PubMed]

21. Hoshizaki, T.B.; Post, A.; Oeur, R.A.; Brien, S.E. Current and future concepts in helmet and sports injury prevention. *Neurosurgery* **2014**, *75*, S136–S148. [CrossRef] [PubMed]

22. Vanden Bosche, K.; Mosleh, Y.; Depreitere, B.; Vander Sloten, J.; Verpoest, I.; Ivens, J. Anisotropic Polyethersulfone Foam for Bicycle Helmet Liners to Reduce Rotational Acceleration during Oblique Impact. *Proc. Inst. Mech. Eng. Part H* **2017**, *23*, 1–11. [CrossRef] [PubMed]

23. Honarmandi, P.; Sadegh, A.M. Modeling and impact analysis of football helmets: Toward mitigating mTBI. In Proceedings of the ASME 2012 International Mechanical Engineering Congress and Exposition, Houston, TX, USA, 9–15 November 2012; American Society of Mechanical Engineers: New York, NY, USA, 2012; pp. 883–891.

24. Von Holst, H.; Halldin, P. Protective Helmet. US6658671B1, 21 December 1999.

25. Kleiven, S.; Hardy, W.N. Correlation of an FE Model of the Human Head with Local Brain Motion–Consequences for Injury Prediction. *Stapp Car Crash J.* **2002**, *46*, 123–144. [PubMed]

26. Kleiven, S. Influence of impact direction on the human head in prediction of subdural hematoma. *J. Neurotrauma* **2002**, *20*, 365–379. [CrossRef] [PubMed]

27. Kleiven, S. Evaluation of head injury criteria using a finite element model validated against experiments on localized brain motion, intracerebral acceleration, and intracranial pressure. *Int. J. Crashworth.* **2006**, *11*, 65–79. [CrossRef]

28. Casson, I.R.; Viano, D.C.; Powell, J.W.; Pellman, E.J. Twelve years of national football league concussion data. *Sports Health* **2010**, *2*, 471–483. [CrossRef] [PubMed]

29. Ekeland, A.; Rødven, A.; Heir, S. *Injury Trends in Recreational Skiers and Boarders in the 16-Year Period 1996–2012*; Snow Sports Trauma and Safety; Springer: Cham, Switzerland, 2017; pp. 3–16.

30. Dickson, T.J.; Trathen, S.; Terwiel, F.A.; Waddington, G.; Adams, R. Head injury trends and helmet use in skiers and snowboarders in Western Canada, 2008–2009 to 2012–2013: An ecological study. *Scand. J. Med. Sci. Sports* **2017**, *27*, 236–244. [CrossRef] [PubMed]

31. Duncan, O.; Foster, L.; Senior, T.; Allen, T.; Alderson, A. A Comparison of Novel and Conventional Fabrication Methods for Auxetic Foams for Sports Safety Applications. *Procedia Eng.* **2016**, *147*, 384–389. [CrossRef]

32. Allen, T.; Duncan, O.; Foster, L.; Senior, T.; Zampieri, D.; Edeh, V.; Alderson, A. *Auxetic Foam for Snow-Sport Safety Devices*; Snow Sports Trauma and Safety; Springer: Cham, Switzerland, 2017; pp. 145–159.

33. Lisiecki, J.; Błazejewicz, T.; Kłysz, S.; Gmurczyk, G.; Reymer, P.; Mikułowski, G. Tests of polyurethane foams with negative Poisson's ratio. *Phys. Status Solidi B* **2013**, *250*, 1988–1995. [CrossRef]

34. Duncan, O.; Allen, T.; Foster, L.; Senior, T.; Alderson, A. Fabrication, characterisation and modelling of uniform and gradient auxetic foam sheets. *Acta Mater.* **2017**, *126*, 426–437. [CrossRef]

35. Alderson, A.; Alderson, K.L.; Davies, P.J.; Smart, G.M. The Effects of Processing on the Topology and Mechanical Properties of Negative Poisson's Ratio Foams. In Proceedings of the ASME 2005 International Mechanical Engineering Congress and Exposition, American Society of Mechanical Engineers, Orlando, FL, USA, 5–11 November 2005; pp. 503–510. [CrossRef]

36. Lakes, R. Foam structures with a negative Poisson's ratio. *Science* **1987**, *235*, 1038–1040. [CrossRef] [PubMed]

37. Gibson, L.; Ashby, M. The mechanics of foams: Basic results. In *Cellular Solids: Structure and Properties*; Cambridge University Press: Cambridge, UK, 1997; pp. 175–234. [CrossRef]
38. Allen, T.; Martinello, N.; Zampieri, D.; Hewage, T.; Senior, T.; Foster, L.; Alderson, A. Auxetic foams for sport safety applications. *Procedia Eng.* **2015**, *112*, 104–109. [CrossRef]
39. Scarpa, F.; Pastorino, P.; Garelli, A.; Patsias, S.; Ruzzene, M. Auxetic compliant flexible PU foams: Static and dynamic properties. *Phys. Status Solidi B* **2005**, *242*, 681–694. [CrossRef]
40. Critchley, R.; Corni, I.; Wharton, J.A.; Walsh, F.C.; Wood, R.J.; Stokes, K.R. A review of the manufacture, mechanical properties and potential applications of auxetic foams. *Phys. Status Solidi B* **2013**, *250*, 1963–1982. [CrossRef]
41. Sanami, M.; Ravirala, N.; Alderson, K.; Alderson, A. Auxetic materials for sports applications. *Procedia Eng.* **2014**, *72*, 453–458. [CrossRef]
42. Allen, T.; Hewage, T.; Newton-Mann, C.; Wang, W.; Duncan, O.; Alderson, A. Fabrication of Auxetic Foam Sheets for Sports Applications. *Phys. Status Solidi B* **2017**, *254*. [CrossRef]
43. Duncan, O.; Foster, L.; Senior, T.; Alderson, A.; Allen, T. Quasi-static characterisation and impact testing of auxetic foam for sports safety applications. *Smart Mater. Struct.* **2016**, *25*, 054014. [CrossRef]
44. Allen, T.; Shepherd, J.; Hewage, T.A.M.; Senior, T.; Foster, L.; Alderson, A. Low-kinetic energy impact response of auxetic and conventional open-cell polyurethane foams. *Phys. Status Solidi B* **2015**, *252*, 1631–1639. [CrossRef]
45. Ge, C. A comparative study between felted and triaxial compressed polymer foams on cushion performance. *J. Cell. Plast.* **2013**, *49*, 521–533. [CrossRef]
46. Martz, E.O.; Lee, T.; Lakes, R.S.; Goel, V.K.; Park, J.B. Re-entrant transformation methods in closed cell foams. *Cell. Polym.* **1996**, *15*, 229–249.
47. U.S. Department of Transportation. *Federal Motor Vehicle Safety Standards and Regulations*; U.S. Department of Transportation: Washington, DC, USA, 1999; 208E.
48. NOCSAE DOC. *022-10m12: Standard Performance Specification for Newly Manufactured Baseball/Softball Batter's*; NOCSAE Standards Database; National Operating Committee on Standards for Athletic Equipment: Overland Park, KS, USA, 2014.
49. Duncan, O.; Allen, T.; Foster, L.; Gatt, R.; Grima, J.N.; Alderson, A. Controlling Density and Modulus in Auxetic Foam Fabrications—Implications for Impact and Indentation Testing. *Proceedings* **2018**, *2*, 250. [CrossRef]
50. Critchley, R.; Corni, I.; Wharton, J.A.; Walsh, F.C.; Wood, R.J.; Stokes, K.R. The Preparation of Auxetic Foams by Three-Dimensional Printing and Their Characteristics. *Adv. Eng. Mater.* **2013**, *15*, 980–985. [CrossRef]

 *applied sciences*

Review

Review of Auxetic Materials for Sports Applications: Expanding Options in Comfort and Protection

Olly Duncan [1], Todd Shepherd [2], Charlotte Moroney [3], Leon Foster [4], Praburaj D. Venkatraman [3], Keith Winwood [2,5], Tom Allen [2] and Andrew Alderson [1,*]

[1]   Materials and Engineering Research Institute, Sheffield Hallam University, Sheffield S1 1WB, UK; Oliver.H.Duncan@student.shu.ac.uk

[2]   Sports Engineering Research Team, Manchester Metropolitan University, Manchester M1 5GD, UK; todd.shepherd@stu.mmu.ac.uk (T.S.); k.winwood@mmu.ac.uk (K.W.); T.Allen@mmu.ac.uk (T.A.)

[3]   Manchester Fashion Institute, Manchester Metropolitan, Manchester M15 6BH, UK; charlotte.m.moroney@stu.mmu.ac.uk (C.M.); p.venkatraman@mmu.ac.uk (P.D.V.)

[4]   Centre for Sports Engineering Research, Sheffield Hallam University, Sheffield S1 1WB, UK; l.i.foster@shu.ac.uk

[5]   School of Healthcare Science, Manchester Metropolitan University, Manchester M1 5GD, UK

*   Correspondence: a.alderson@shu.ac.uk; Tel.: +44-(0)-114-225-3523

Received: 10 May 2018; Accepted: 31 May 2018; Published: 6 June 2018

Abstract: Following high profile, life changing long term mental illnesses and fatalities in sports such as skiing, cricket and American football—sports injuries feature regularly in national and international news. A mismatch between equipment certification tests, user expectations and infield falls and collisions is thought to affect risk perception, increasing the prevalence and severity of injuries. Auxetic foams, structures and textiles have been suggested for application to sporting goods, particularly protective equipment, due to their unique form-fitting deformation and curvature, high energy absorption and high indentation resistance. The purpose of this critical review is to communicate how auxetics could be useful to sports equipment (with a focus on injury prevention), and clearly lay out the steps required to realise their expected benefits. Initial overviews of auxetic materials and sporting protective equipment are followed by a description of common auxetic materials and structures, and how to produce them in foams, textiles and Additively Manufactured structures. Beneficial characteristics, limitations and commercial prospects are discussed, leading to a consideration of possible further work required to realise potential uses (such as in personal protective equipment and highly conformable garments).

Keywords: injury; impact; indentation; comfort; protective equipment; negative Poisson's ratio; foam; textiles; Additive Manufacturing; finite element modelling; auxetic

---

1. Introduction

Auxetic materials have a negative Poisson's ratio (NPR) [1], meaning they expand laterally in one or more perpendicular direction/s when they are extended axially. Poisson's ratio (v) is the negative of the ratio of lateral to axial strain (Figure 1). The potential application of foams, textiles and additively manufactured (AM) auxetic materials to sporting protective equipment (PE), as well as other forms of impact protection, has been discussed in articles with a focus on materials (e.g., [2–5]) and sport (e.g., [6–8]). Auxetic foam was the first man made auxetic material [9], and makes up a significant proportion of the scientific literature and therefore this review. Foam studies typically use relatively established fabrication methods for auxetic polyurethane (PU) foams, with tests often based upon those outlined in standards used to certify sports safety equipment (i.e., British Standards Institution—BSI, International Organization for Standardization—ISO, American Society for Testing and Materials—ASTM). Idealised structures (e.g., those fabricated via AM) and textiles can be designed to have repeat patterns that provide NPR in either one or two planes [5].

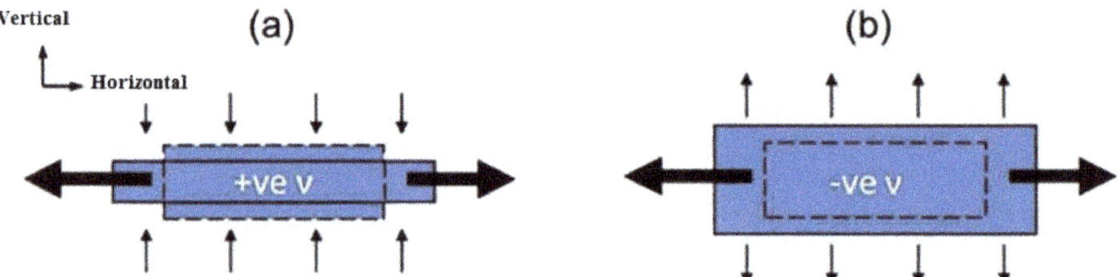

Figure 1. Lateral (vertical) deformation due to Poisson's ratio during tensile axial (horizontal) loading for (**a**) a conventional material and (**b**) an auxetic material. Thick and thin arrows correspond to deformation due to loading and Poisson's ratio respectively.

There are numerous reviews into auxetic materials and/or structures [10–17], but there has not yet been one focussing on their application to sporting goods. Deficiencies in standards (or a lack of a standard in some instances) and PE [18,19] indicate changes in equipment testing and design are likely, with the need for novel and improved materials to meet new requirements.

Recent developments in the fabrication of auxetic PU foam have deepened our understanding of the mechanisms that can fix an imposed cell structure, while improving and diversifying fabrication methods. Developments include rapid fabrications [20,21] and the ability to tailor elastic modulus and Poisson's ratios of anisotropic and gradient foam samples [22,23]. Finite element modelling (FEM) has been applied to auxetic materials [24–26] but has not been fully explored to support their implementation within sports equipment. With developments in AM [27], the creation of idealised auxetic structures (computationally tested using FEM) has become commercially viable [28]. Developing fabrication techniques for auxetic textiles and composites are also bringing both protective and form-fitting auxetic garments closer to realisation [29].

The focus of this review is on sports equipment and auxetic materials, predominantly auxetic foams, textiles and FEM/AM for impact protection. NPR can provide uniquely high indentation resistance [30,31] and fracture toughness [32], lending auxetic materials well to impact force or acceleration attenuating scenarios, regularly demonstrated by impacting auxetic foams [2,3,33]. Auxetics' multi-axial expansion [34,35] and double curvature [9,36,37] could improve comfort, fit and durability in sporting garments and personal protective equipment (PPE). Multi-axial expansion could also be useful in filtration applications [9,38–40], while gradient structures' enhanced bending stiffness [41,42] could reduce the mass of skis, snowboards, tennis rackets or hockey sticks (to name a few), without sacrificing stiffness. An initial summary of current sporting PE and PPE will lead to descriptions of common auxetic structures and materials, how to create them and how they could resolve issues in sporting PE, PPE, garments and other sports equipment.

2. Introduction to Sporting Protective Equipment

Injuries in sport are common and place a significant burden upon participants and national economies [43,44], estimated at $525 (~£380) million per year in The Netherlands [45] for example. The main methods of intervention are elimination, modification (a reduction in severity or likelihood of injury) and reaction (i.e., medical) [44]. Preventative measures such as sports safety equipment, and/or rule changes, are more cost effective than reactive procedures [44] and successful products can increase a manufacturer's share of the sporting goods market (~~$90/£66 billion in the USA in 2017 [46]). Sporting PE is intended to reduce risks and is typically either an addition to the playing field or environment (i.e., crash matt or barrier, known herein as PE) or PPE. Both equipment types provide; (i) protection against impact, dissipating and absorbing energy while reducing peak forces/accelerations, pressures and in some cases such as helmets, impulses [43,44,47–49] and (ii) protection from penetration, abrasions and lacerations [18,47,48,50]. Sporting PPE can also provide support to joints, muscles and the skeleton [18,44,47,48,51].

Sporting PPE (Figure 2a) often contains a shell—typically a stiff material or non-Newtonian fluid layer—to spread forces and reduce pressures [47,48,50,52]. Impacts are typically attenuated by elastic [44,52], viscoelastic and/or permanent deformation (i.e., crushable foam in cycling helmets) of a cushioning material with a lower compressive stiffness than the shell [47,48,50,52]. Visco-elastic and permanent deformation reduce impulses, improving the level of protection [44].

Sporting PPE must sometimes cover large areas and should ideally be low in mass and bulk to reduce restriction of movement, fatigue and heat build-up. Foam is typically used for the cushioning layer in PPE, and with limited thickness to compress and decelerate impacting bodies material selection is crucial. In contrast, PE such as crash mats (where bulk, mass and ergonomics are less critical) are often large and thick, allowing more gradual deceleration over a greater distance through compression of relatively compliant foam [52]. For both sporting PE and PPE, deceleration (or force) ideally reaches a maximum yet safe value at low strains, before plateauing and deforming with no additional load to safely maximise energy absorption (the integral of decelerating force with respect to deformation) before foam densification (at high strains) causes high deceleration (or force) [53]. Crash mats are used in a variety of conditions; protecting skiers, snowboarders and mountain bikers (who can travel at high speeds) from hazards on a mountain (i.e., lift poles or trees), often in extreme and variable climates, to providing padding to gymnasts on flat surfaces at room temperature.

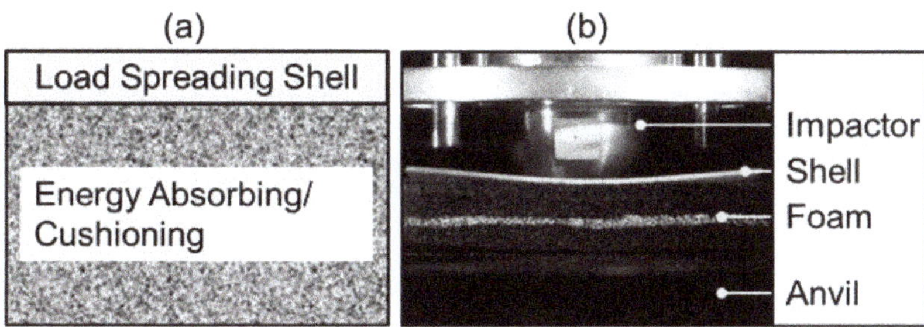

Figure 2. (**a**) Schematic of a typical piece of Personal Protective Equipment containing pressure dissipating material (hard outer shell) and energy absorbing/cushioning material (deformable foam); (**b**) Impact adapted from British Standard 6183-3:2000.

Sport safety equipment is often certified according to tests specified in standards and regulations (i.e., [54–58]), which typically specify a maximum allowable peak force/acceleration under impact (i.e., Figure 2b). To perform well in these tests, a product should absorb or dissipate as much energy as possible to prevent force/acceleration from passing its allowable limit [59]. Criteria within standards (such as impact energies, velocities and masses) are not always well justified and tests typically feature fixed rigid anvils rather than biofidelic (human like) surrogates, meaning they are not overly representative of the scenarios where the equipment may be required to perform [44]. In some cases there is no standard, so manufacturers certify products against a standard for another safety device, which can be misleading for consumers. McIntosh discusses the increased chance of injury when perceived protection offered by equipment is incorrect [19].

Crash mats are typically certified to BS EN 12503-1:2013 [60] as intended for gymnasium use, but include impact tests for outdoor activities such as pole vaulting. BS EN 12503-1:2013 is not reflective of the sometimes harsh and variable outdoor environment (weather conditions, surfaces, etc.) where these mats can be located, and tested parameters do not reflect realistic impacts [58]. There are other occasions where specific standards are not available, and so a proxy is used. Two examples from snow sports are wrist protectors, certified to EN 14120:2003 [61] for roller sports [51], and back protectors, which are often certified to EN1621 (motorbike) [62]. Schmitt et al. found that snow sport participants expect back protectors to protect the spine, but EN1621 does not test against scenarios likely to cause spinal injury [18]. Inspection of PPE (impact shorts, knee guards, elbow guards and back protectors etc.) marketed for adventure sports including mountain biking, kayaking and snow sports shows EN1621 [62] & EN 14120:2003 [61] (for motor-biking) are repeatedly used in place of a dedicated standard.

Some standards have received criticism for not providing adequate tests even when applied to their intended field. Reaction to a number of high profile deaths and serious injuries as well as multiple awareness campaigns [63] raised the issue of helmet use in snow sports considerably, but head injury rates have remained fairly constant [63,64]. Scandals in the National Football League (NFL) culminated in high proportions of Chronic Traumatic Encephalopathy (CTE, up to 99% in a bias but large sample set) that could have contributed to early death, suicide and dementia in players [65].

In some cases, equipment and regulations intended to protect sports people are clearly unsatisfactory. As an example of equipment not meeting expectations, attenuation of rotational acceleration is thought to be critical in protection from concussion in sports [66,67]. Standards, however, typically assess helmets based upon their attenuation of linear accelerations (e.g., EN1077 & ASTM F2040, & F1446 [54,56,68]) and resistance to penetration (e.g., EN1077 [56]) based upon direct impacts [69,70]. Standards can be updated or replaced, for example BSI 6685-1985 for motorised vehicle helmets (a previous revision of Reference [71]) replaced BS2495:1977 and BS5361:1976 to include oblique impacts. The standard for cricket helmets (BS7928:1998) was amended (BS7928:2013) to include impacts by cricket balls [72] following findings that cricket helmets were not sufficiently attenuating acceleration under high speed impacts [73]. Commuter cycling (where cyclists travel alongside motor vehicles) is becoming increasingly popular due to clear benefits to health, congestion and emission levels, as well as improved facilities such as dedicated lanes. Safety concerns are a major barrier to participation in commuter cycling [74,75] but helmets are still only certified to protect from linear accelerations [76].

Sports equipment is a competitive, rapid uptake market. Manufacturers search for new technology to remain competitive, achieve the highest possible levels of certification and improve safety. One approach to solve the problem of rotational acceleration in helmets is a slip plane between the shell and crushable foam [77]. Slip plane technology is included in some commercial helmets, despite a lack of experimental evidence to support a reduction in concussion risk [78–81], highlighting the problem caused by insufficient standards. As an example of material development in sporting PPE, trends over the past twenty years have favoured lightweight, ergonomic equipment which does not sacrifice performance in standard tests [49]. Non-Newtonian fluids with high energy absorption were developed and can act as both a shell and cushion [82,83], offering comfort and flexibility under normal use and increased stiffness under impact. Non-Newtonian materials in isolation can pass certification tests for sporting PE and PPE [18,44,47,48,50]. Scientific literature highlights limitations in standards, as well as associated certified products including helmets [69,79–81], back protectors [18,50] and wrist protectors [51,84,85]. Recent trends look to include the use of more representative surrogates rather than rigid anvils [84,86,87] and tests designed for specific sports [50,51] to replace proxy standards (e.g., [61]). Solutions including novel materials are needed to reduce the effect of sports injuries and meet required improvements to standards (i.e., as per BSI 6685-1985 [71] and BS7928:2013 [72]).

3. Common Auxetic Materials and Structures

Auxetic research began in earnest with open cell foam [9], the first example of a man made NPR material. Subsequently, auxetic materials have now been developed or discovered in other nanocrystalline [88] and microporous [89]) polymer, ceramic [90], metallic [91] and composite [92] forms, and include natural systems [93]. Auxetic materials such as foams, textiles and AM structures are often classed as mechanical metamaterials; with unexpected macro-scalar (effective) characteristics caused by their micro/nano-structure [15].

This section will first discuss the modelling of auxetic structures and mechanisms, before considering auxetic foam fabrication, properties and characteristics, followed by some of the other auxetic materials of relevance including textiles and those manufactured by AM.

3.1. Modelling Auxetic Materials and Structures

Numerical [94–97], molecular [98] and FE models [99–101] have all been applied to auxetic materials. These models, used individually or combined, typically analyse the elastic deformation mechanisms of auxetic materials. The most common microstructures of cellular auxetics can be designated into three types; re-entrant, chiral and rotating units (Figure 3a–c). The 2D re-entrant structure (such as Figure 3a,d) was the first to be modelled numerically [95] and was developed in early work on auxetic honeycombs and foams. Other early models included a mechanical model of isotropic 2D/3D frameworks of rods connected by sliding collars [102] and a thermodynamic model of a 2D assembly of hard cyclic hexamer 'molecules' [103]. Chiral auxetics (Figure 3b)—asymmetric structures that are non-superimposable on their mirror image [104]—achieve NPR through cooperative node rotation-induced bending of connecting ligaments. Rigid rotating units (Figure 3c) can have NPR, dependent on the rotation of connected squares [105] or other shapes [106–108], and this model has been used as an alternative to the re-entrant model in auxetic foams [109]. Other models and structures exist, including the missing rib model [110], cellular systems featuring pre-defined mechanical instabilities [111] and the nodule-fibril model for polymers [112]. Auxetic behaviour induced by elastic instability includes systems consisting of 2D tessellation of elliptical [113] and other shape [114,115] voids and 3D tessellations of holes [116]. Non-porous sheets with tessellations of spherical dimples have also been investigated for auxetic response [117].

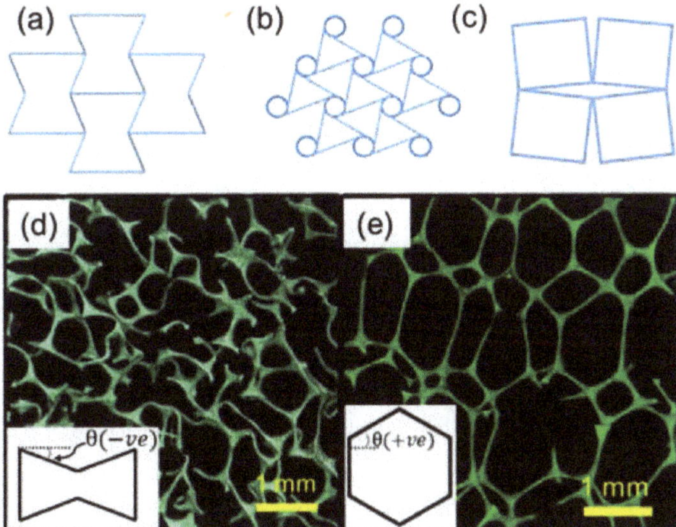

Figure 3. (**a**) Assembled 2D re-entrant structure; (**b**) 2D Chiral structure; (**c**) 2D rotating squares structure; (**d**) Micro Computed Tomography of polyurethane auxetic open cell foam (depth = 1 mm), pop-out showing simplified 2D diagram of a re-entrant cell, θ = re-entrant angle (negative value for angle below horizontal axes); (**e**) Micro CT scan of conventional open cell PU foam (depth = 1 mm), with pop-out showing simplified 2D diagram of a conventional polyhedral cell, θ = cell rib angle (positive value for angle above horizontal axes).

Re-entrant models have been used to investigate the deformation mechanics of cellular auxetics. Deformation due to flexure of the cell wall alone was initially considered, with the cell ribs assumed to behave as beams of uniform thickness [118]. Combined hinging (change of angle between ribs) and stretching (increase in rib length) of the cell wall was later modelled [98], as well as a 14-sided polyhedron, specifically for open cell foam, with cell rib bending as the main deformation mechanism [119]. All three deformation modes have been modelled simultaneously [96,120], confirming that the change of cell parameters (e.g., cell length or rib angle) directly affected Poisson's ratio and Young's modulus. The combined hinging, stretching and bending model [96] has been adapted to accurately predict the mechanical properties of gradient, isotropic and anisotropic auxetic foam [23].

Auxetic application in other fields has progressed with the use of FEM, such as auxetic cores used in sandwich plates that reduce shear stresses [121], anti-tetrachiral stents demonstrating desired radial expansion with axial stability [122] and smart auxetic honeycombs in structural health monitoring [123]. FEM can predict the effect of changing design parameters of auxetic structures, such as cell wall thickness [97,124], unit cell cross-sectional area [94], relative stiffness of modelled beams [125] and ratio of vertical to oblique cell wall length [126]. Simulation results could therefore aid and inform the design of future prototypes—e.g., to reduce a structure's mass while maintaining the maximum load it can support. FEM has also been used to develop understanding of the effect of plastic deformation on auxetic behaviour [127,128] and generate specific material models to explore impact resistance of auxetic honeycomb structures [129]. A framework for modelling auxetic foam using the continuum approach at large strains has been developed with FEM [130–132] and could be extended to model composite structures made from auxetic materials.

Several FEM studies established that the angle of re-entrant cell ribs effects Poisson's ratio [26,133–135], demonstrating cell parameter manipulation. FEM has also highlighted that other internal cell parameters (e.g., length, height, thickness) can be changed to tune the Poisson's ratio, and that flaws in the re-entrant cell structure can reduce NPR's magnitude towards zero [136]. Innovations are facilitated by FEM, such as the combination of two auxetic structures within one design (re-entrant hexagon and arrowhead) [137], honeycombs with ribs made of 2-phase composite material [138], the development of curved honeycomb unit cell designs [139], the creation of auxetic honeycombs inspired by the spider's web [140] and the formation of auxetic systems made out of readily available materials [141] and recycled rubber [142].

FEM has also been employed to model auxetic two-phase systems without voids by filling the spaces of a cellular first phase with a matrix second phase. This was an idea first postulated in 1992 and modelled analytically for elastically isotropic laminates by Milton [143] and using FEM for 'network embedded composites' by Evans and co-workers [144]. More recently, FEM has been extended to a two-phase core employing conventional and/or auxetic phases for out-of-plane auxetic behaviour [141], and to sandwich panel composites with the two-phase core employing established auxetic cellular structures and/or auxetic matrix [145–147], as well as structures developed using topology optimisation routines [145,148]. The ability to have temperature-dependent auxetic behaviour in such systems has also been investigated using FEM by allowing one or both phases to possess temperature varying Young's modulus [149,150].

The effect of design changes on Young's modulus and Poisson's ratio can be adjusted and improved before manufacture [121], and FEM can validate analytical results [151–158]. Full validation of FEM often requires comparison and agreement with experimental testing [159–167], which can potentially lead to extrapolation or tailoring of models for specific applications. Experimental test data can be input into FEM solvers to validate and develop material models [168]. With a validated model, design changes can be implemented and investigated without numerous iterations of manufacturing and testing. FEM can then be used alongside optimisation tools to further improve auxetic designs [169–173], although the 'modelling theory' remains to be examined stringently [174].

Dynamic effects can also be examined with explicit FEM solvers. The energy absorption of a re-entrant structure when crushed by a rigid wall has been analysed [175], but not experimentally validated. Analytical formulae were however used to validate FEM suggestions that auxetic honeycombs have higher energy absorption than conventional honeycombs in crushing strength analysis [176]. When varying the impact energy on a jounce bumper [27,177], FEM closely matched experimental values for the material's Poisson's ratio and compressive modulus and the jounce bumper's force and displacement. A subsequent auxetic bumper model showed less vibration in compression than a non-auxetic equivalent. A range of (FEM) parametric studies compared rate-dependent blast resistance performance in military applications [178,179]. Auxetic sandwich panels had higher impact resistance than their conventional counterparts, validated with comparison to analytical models rather than experimental data. Sporting PE and PPE typically aims to minimise harmful impact forces/accelerations through high strain rate compression of energy absorbing materials [180,181], and explicit solvers have previously facilitated such investigation [47,48,182].

3.2. Fabricating and Characterising Auxetic Foam

Auxetic foam is often fabricated to experimentally show expected enhancements (i.e., to impact force attenuation or vibration damping) due to NPR [2,3,33,183]. Auxetic foam fabrications typically change the cell structure of open cell foam (referred to herein as the parent foam) to give it an NPR. Lakes' thermo-mechanical fabrication method [9] first applies a volumetric compression ratio (VCR, normally defined as original/final volume and typically between 2 and 5 [184]) to a parent foam in all three orthogonal axes to buckle cell ribs [118]. The applied compression changes the cell shape and causes the re-entrant structure, as cell ribs buckle beyond ~5% compression [118]. The compressed foam is then heated to a set temperature to encourage permanent plastic deformation. The temperature is typically between 130 and 220 °C [13], often referred to as the 'softening temperature' [20,185]. Finally, the foam is cooled to fix the imposed structure [9]. Buckling the originally straight ribs (Figure 3e) gives the polyhedral cells a re-entrant, contorted cell structure [9] (Figure 3a,d). Typical sizes were initially small, in the order of 20 × 20 × 60 mm (following fabrication), although larger 'scaled up' samples have subsequently been fabricated (e.g., [185–188]).

Fabrication methods for auxetic foams have developed and diversified since Lakes' initial study [9]. Stretching samples after cooling was introduced to reduce adhesion between cell ribs and residual stresses [189]. More recently, fabrication processes have been split into stages including; multiple heating cycles (with foam removed from the mould and stretched by hand in between to reduce residual stresses, flaws and creases [190]) and the addition of an annealing stage—heating below the softening temperature. Foams are typically annealed at 100 °C [37,185,190], but alternatives to annealing include slow cooling in the mould in air [20,191] or cooling outside the mould in air [192]. In an attempt to reduce creases and flaws when inserting foam into the mould, olive or vegetable oil and WD-40 have been used to lubricate moulds [185,189,193].

Another line of investigation has attempted to increase the range of Poisson's ratios and Young's moduli achievable following thermo-mechanical fabrications. The heat, time and compression applied during fabrication can be adjusted to give higher magnitude NPR and increased stiffness [129]. Heating for longer or at a higher temperature ('over-heating' while using typical compression levels) gives a positive or near zero Poisson's ratio re-entrant foam [30,194]. Over-heated re-entrant samples have comparable density to typical auxetic foams and a linear stress strain curve without the presence of a plateau region [30,194]. Comparing auxetic and over-heated foam with a positive Poisson's ratio [194] could demonstrate the effect of Poisson's ratio on characteristics such as vibration damping or impact force attenuation.

In depth analysis shows the complexity and diversity of chemical constitution in polymers (including those found in foams) [20]. Polymeric microstructure [20], and microstructural changes caused by heating [195] have only recently been investigated in relation to auxetic foam fabrication. Li & Zeng targeted styrene acrylonitrile copolymer (SAN) particle bonding, which has a glass transition temperature of around 110 °C. Such, or other, electron donating groups [15] can provide fixing mechanisms for imposed structures.

Sporting products often include foam sections which have a larger planar area than typical auxetic foam samples (~20 × 20 × 60 mm) [9,185,196], and numerous scaled up fabrications have been attempted [4,7,23,185–187]. Large fabricated samples often exhibit random and ordered inhomogeneity. Random inhomogeneity is caused by flaws such as surface creasing [185], due to difficulty compressing large samples of foam into the mould [186]. Ordered inhomogeneity was observed in fabricated cubes (150 mm sides), where the centre had the lowest density [4,8]. Possible explanations include reduced compression towards the centre of the cube during fabrication or a thermal gradient that meant the internal structure was not sufficiently fixed and re-expanded following fabrication. The specific influence of temperature gradients and compression gradients during auxetic foam fabrication is unclear.

Solutions to increase homogeneity when fabricating large samples of foam include; (i) a mould with moveable walls which can be assembled around the foam and then used to apply compression [186,197]; (ii) a multi-stage compression process with an intermediate VCR to reduce insertion forces [185] and (iii) a vacuum bag to apply compression [187]. Vacuum bags can apply consistent pressure through thickness, reducing external densification and allowing fabrication of additional shapes (i.e., curved) [187]. Recent work utilising rods passing through large sheets of foam [7,8,23,188] is the only published method which seeks to apply controlled distribution of material within the mould. 'Felted foams' are fabricated commercially by compressing open cell foam between two heated plates to impart an anisotropic, re-entrant cell structure and direction dependent NPR [33].

Small scale fabrication parameters depend on the type of foam used and the compression levels applied [20,185]. It is likely that changes to fabrication parameters may be needed when 'scaling up' fabrications, and there may not be a 'one size fits all' style best approach. A comparative evaluation of gradual compression, vacuum bag compression and controlling internal compression using rods would, nonetheless, highlight advantages and disadvantages and aid selection and adaptation of the most appropriate method for fabricating large samples of auxetic foam.

Alternatives to the thermo-mechanical fabrication process [9] include using a solvent or softening agent instead of heating. Acetone [196] and pressurised CO_2 [21] can be used, depending on polymeric constitution and types of bonding present within the original foam. Softening methods can be combined, and both acetone [2] and CO_2 [21] have been used in combination with heat. Chemicals can be used to target specific bonds (i.e., SAN particle bonds) [20]. Liquid solvents (such as acetone) will require a drying phase and so some form of gradient is likely in larger samples. Provided there is no significant gradient due to diffusion of gases from the centre of samples during/after fabrication, the CO_2 softening route could reduce the effect of temperature gradients in thermo-mechanical fabrications.

Attempts to apply thermo-mechanical fabrication methods typically used for open cell foam to closed cell foam can rupture the foam's cell walls [185,198]. NPR has still been achieved along one axis in closed cell low density polyethylene (LDPE) foam, by combining thermal softening (at 110 °C) and high hydrostatic pressure (662 kPa, applied by a pressure vessel) over 10 h and maintaining the pressure for a further 6 h after cooling [199]. Heating for an hour at 86 °C, then subjecting to vacuum pressure for 5 min prior to sudden restoration of atmospheric pressure also produced uniaxial NPR in the same LDPE closed cell foam [199]. Slow diffusion of gas through cell walls was similar to predicted values, suggesting the cell walls in the auxetic LDPE foam had remained intact. Solid state foaming (sticking together pieces of closed cell foam cut in a re-entrant shape) [200], syntactic foam processes (embedding degradable/collapsible beads with a re-entrant shape into a molten/liquid polymer) [201] and AM [202] have also produced auxetic foam-like structures.

Measuring strain in compliant, often inhomogeneous foam or foam like structures requires non-contact methods. Wide ranging studies, employing a variety of test protocols, have been undertaken to characterise auxetic polymeric foams for structural, mechanical, thermal, filtration and impact properties, for example. Isotropic auxetic foams with Poisson's ratio between 0 and −0.7 have been reported (e.g., [3,9,20,22,23,189,195,203]). Higher magnitudes of NPR (<−1) have also been reported for anisotropic auxetic foams [187,190]. Density measurements of whole or dissected samples are useful in assessing the extent of, and variation in, volumetric compression throughout thermo-mechanically fabricated re-entrant foam [22,204]. Standardisation while characterising samples would help assess levels of agreement between different studies. Where possible future testing should be undertaken in accordance with, or based upon, the appropriate ASTM standard (e.g., ASTM-D412—15a [205]) for quasi-static tensile testing, which requires communication of sample dimensions, strain rate, range and measurement method and joining/contact methods between sample and test rigs.

3.3. Fabricating Auxetic Textiles

Auxetic textiles have the potential to contribute enhanced mechanical properties to textile applications in sport—including for apparel, equipment and injury prevention and treatment. Developments of auxetic fibres and fabrics, and their production methods, have enhanced the potential to use auxetic textiles commercially. Auxetic yarn with a high magnitude of NPR (<−2) can be produced from standard, non-auxetic fibre materials and conventional textile manufacturing processes [206]. Fabrics can, therefore, be manufactured from auxetic yarns without the need to develop new techniques.

The auxetic yarn in Reference [206] comprises a relatively high modulus thin wrap fibre around a lower modulus thick core. Both wrap and core components are typically conventional fibres. When extended the (initially helical) wrap fibre becomes straight and pushes the (initially straight) core fibre into a helix. Since the core is thicker than the wrap, the final stretched yarn is thicker with the helix core than the initial un-stretched yarn with a helical wrap. Due to the double helix nature of the wrap and core construction, the auxetic yarn is referred to in the literature as the double helix yarn (DHY) [207] and also the helical auxetic yarn (HAY) [208–211]. In this review we refer to this yarn as the DHY. Development of the DHY includes an auxetic plied yarn [212], auxetic braided structure [160], and a DHY where the 'wrap' fibre is stitched in place to offer more control over its behaviour by preventing fibre slippage [213]. Heat treatment has also been utilised to solve the issue of slippage [214]. The initial angle at which the wrap fibre is spun around the core fibre, and the diameter ratio of wrap to core fibre influences the value of the NPR, whereby a lower yarn wrap angle increases the magnitude of NPR [215–217]. The DHY offers increased energy absorption under impact [209].

The production of auxetic polymeric monofilament fibres has also been reported and utilises a process of continuous melt extrusion [218,219] to form a microstructure of interconnected surface-melted powder particles. NPR arises at the microscale of these monofilaments, in contrast to conventional filaments, for which mechanical properties, including Poisson's ratio, arise at the molecular (polymer chain) level [220]. The developed auxetic monofilaments have been used to demonstrate enhanced fibre pull-out resistance of auxetic fibres, requiring double the extraction force (than their conventional counterparts) [221], which could enhance the longevity and robustness of sports apparel and equipment.

Auxetic yarns and fibres have been incorporated into fabrics. The DHY has been employed to produce auxetic woven fabrics for composite reinforcement [206] and potentially as medical bandages [217]. Auxetic plied yarns have also been used in woven fabrics where yarns with shorter floating threads promote a greater magnitude of NPR in the fabric [222]. The auxetic monofilament has been used to develop knitted, woven and non-woven fabrics [223].

Auxetic fabrics can also be produced from conventional fibres. Out-of-plane NPR has been induced in non-woven fabrics, produced thermo-mechanically [224,225], via compression between two flat, heated plates to tilt and buckle needle punched fibre bundles. The through-the-thickness auxetic samples consist of bent fibre bundles perpendicular to the surface of the fabric. Under tension, the bent fibre bundle becomes re-orientated, in turn pushing on surrounding fibres and causing the fabric thickness to increase. More recently, auxetic foldable woven fabrics have been fabricated through the use of conventional yarns and weaving technologies [226].

Knitting technologies offer the benefit of a high structure variety [29], enabling the fabrication of a range of different geometrical auxetic textile patterns. Auxetic fabrics can be manufactured using conventional yarns on commercial warp knitting machinery, which can be produced at rates and quantities comparable to non-auxetic warp knits [227]. Hexagonal structures are used for NPR warp-knitted textiles, including a rotational hexagonal structure [228] and re-entrant hexagonal net structures [229]. These hexagonal meshes have since influenced the fabrication of in-plane auxetic spacer fabrics (Figure 4a) [230,231]. Auxetic spacers exhibit excellent shape fitting to complex curves [232], high energy absorption and indentation resistance, and these properties could be enhanced with a greater magnitude of NPR [233]. Yarn loading capacity in NPR spacers affects tensile energy absorption [234], and FEM has successfully predicted the tensile NPR of a warp-knit spacer fabric [166]. Further research and tailoring of NPR spacers could aid the fit of sporting PPE that covers dome-like surfaces, such as protective head-wear [232].

Auxetic weft-knitted fabrics have also been fabricated with computerised flat knitting machines, based on three geometrical structures; (i) foldable structures (Figure 4b); (ii) rotating rectangles and (iii) re-entrant hexagons [235]. When using conventional yarns, the fibre type and geometrical structure are critical to the NPR of these fabrics and they can be adapted to tailor their Poisson's ratio and elastic modulus. Large NPRs (<-1) are possible, depending on the structure and yarns employed [235]. Auxetic foldable weft-knitted textiles have also been used as the reinforcement in polymeric composite materials [236]. Developments of auxetic weft-knitted fabrics will find a wide variety of potential applications in different fields, such as PPE for sportswear and headwear, but more research needs to be undertaken to explore these potential applications further [29].

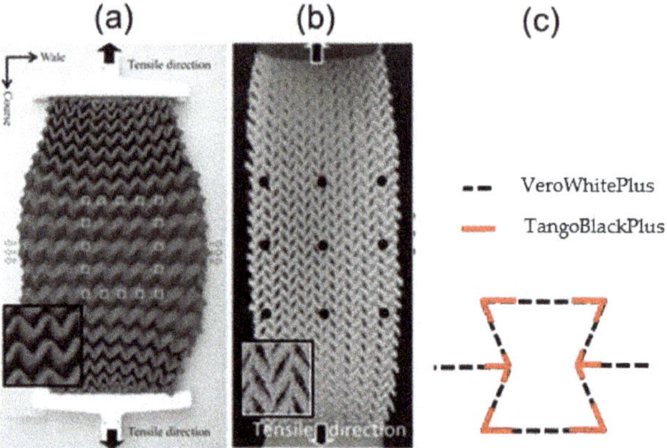

Figure 4. (**a**) Points marked on auxetic warp-knit fabric under tension [36]; (**b**) NPR effect of a weft knit fabric subject to extension [29]; (**c**) Dual-material auxetic metamaterial.

3D auxetic textiles have been developed for composite reinforcement applications [5,237]) through the use of warp and weft yarns [238] with stitch yarns [239] and auxetic fibres [240]. A four layer woven 3D auxetic textile structure has been used as composite reinforcement through a vacuum-assisted resin transfer moulding process [241]. Charpy impact tests (ASTM D6110-17 [242]) found the energy absorption of the auxetic textile was 6.7% higher than that of its conventional counterpart [241]. 3D Auxetic textiles have also been employed to reinforce a foam matrix fabricated by injecting and foaming [243,244]. With through-thickness NPR, 3D textiles may soon provide the enhanced protection associated with NPR materials to sports apparel and PPE [5].

There has, then, been a significant increase in the development and variety of auxetic fabrics since 2011. Due to the range of auxetic textiles under development and their unusual properties, these materials have a wide variety of potential applications [14]. For sports apparel, auxetic textiles could enhance comfort and performance in PPE, with the unique ability to fit easily to curved shapes combined with high compressive energy absorption and indentation resistance [245]. In order to implement auxetic textiles commercially, a future focus should be given to factors affecting their usage and wearability [246]. To embed or incorporate auxetic materials into sports garments, they should be relatively soft and flexible, lightweight, breathable and mouldable or available in a variety of shapes and thicknesses (5 to 10 mm). They may also be required to be homogeneous and/or joinable using flat lock seams [83].

3.4. Additively Manufactured Auxetic Materials and Structures

As alternatives to often inhomogeneous foams or textiles, the creation of auxetic materials with repeatable structures using various AM techniques has been investigated (e.g., fused deposition modelling [137], selective electron beam melting [175], selective laser sintering [247] and lithography-based ceramic manufacturing [248]). AM is a natural extension of FEM, as the existing modelling file can be converted to be compatible with most AM technologies to fabricate the modelled structure [167,249–252]. The AM structure can be experimentally tested to validate and potentially improve FEM [162,253–255], and key parameters can consequently be tuned and investigated with a particular application in mind.

Example materials used for AM in auxetic research include compliant rubbers and plastics. Metals and ceramics can also be made via AM but these studies have been omitted due to their incompatibility with sporting PE and PPE. Auxetic 3D chiral lattices have been modelled and produced by AM using TangoBlack® (Stratasys, Eden Prairie, MN, USA), a rubber-like AM filament [256]. AM has also combined two materials of different stiffness within; (i) one re-entrant unit cell [202,257,258] (Figure 4c); and (ii) novel chiral auxetic structures consisting of four 'base' unit cells surrounding a smaller 'core' chiral unit cell [259]. Using two materials of different stiffness provides a designer with greater control over unit cell deformation. The chiral auxetic has since been developed to use a re-entrant cell as the 'core' unit cell [260], meaning softer hinges are not necessary and the structure can be made from one filament material. The location and amount of each dual material provides additional ways to tune the mechanical properties (e.g., stiffness) of a structure without changing its geometry; as illustrated with the FEM of layered auxetic plates [261]. FEM and AM used in conjunction could facilitate such a parametric study with relative ease, for example by varying repeated structures [262].

AM structure's Poisson's ratio can be tuned by altering the geometry of a unit cell [263–265], enabling the design and analysis of novel auxetic structures using FEM [266,267] to help validate existing analytical formulae. Matlab has also been used to run FEM parametric analysis [167]. NinjaFlex®, (Ninjatek, Manheim, PA, USA) an elastic and flexible thermoplastic PU, has been used in the AM of auxetics [137,268–270], and in impact testing studies [269,270], with FEM used to highlight auxetic structures' desired shock absorption capacity. Elsewhere, impact testing on non-auxetic NinjaFlex® honeycombs demonstrated that increasing strain rates (0.01 to 0.1 s^{-1}) resulted in an increase of energy absorbed (0.01–0.34 J/cm^3) [271]. Both strain rate and energy absorption are key considerations for auxetic material applications in sporting PE and PPE.

3.5. Gradient Materials

Gradient materials can be made from one material by varying its macro-structure to have different mechanical and structural properties in different regions (e.g., auxetic and conventional regions). Gradients can be discrete [22,23,272] or continuous [22,272]. Composite sandwich structures employing discretely gradient honeycombs have a higher compressive modulus and are stronger than conventional sandwich structures [41]. The same sandwich structures exhibit a large increase in bending stiffness at the transition between conventional and NPR regions [42]. Opposing synclastic/anticlastic (domed/saddled) curvature associated with negative/positive Poisson's ratio respectively [9] cause increased shear modulus between regions and a localised stiffening effect. A similar stiffening effect can be seen in AM honeycombs with re-entrant inclusions during tensile tests [273]. Stiffening between discrete gradients of conventional and NPR could increase the bending stiffness (of the whole structure or specific regions) in sports equipment that contains honeycomb or fibre reinforced composites (e.g., skis, snowboards, tennis rackets and hockey sticks to name a few). Auxetic composite sandwich structures have not been tested for sports applications.

Gradient foams have been fabricated, by applying variable compression gradients to different sample sizes (~2 × 2 × 2 cm to ~30 × 30 × 2 cm) using rods [22,23] and/or selecting mismatched uncompressed foam and mould shapes [22,23,272] (e.g., uniform foam sample compressed in a tapered mould). These gradient auxetic foam samples can exhibit vastly different cell structure and mechanical properties in different regions, which can be explained using an analytical model [23]. Gradient structures could be employed to foams or other materials (i.e., fabrics) to allow the development of garments which will fit to the wearer and adapt to their shape as they move.

4. Expected Characteristics and Supporting Evidence

Beneficial characteristics of auxetic materials include increased shear modulus, indentation resistance [10,30], dynamic [274,275] and static compressive energy absorption [8,37] and decreased bulk modulus [10,30]. Increased indentation resistance [30] and compressive energy absorption [2,8,33,37,204] have been shown experimentally in comparisons of auxetic and conventional foams. Lateral expansion due to axial tensile loading makes auxetic structures/materials ideal candidates for straps in apparel, increasing area to spread increasing loads and prevent 'digging in' [276]. Many of the characteristics which provide unique enhancements for auxetic materials come from Elasticity theory [10,277].

For isotropic materials experiencing elastic deformation, Young's modulus (*E*) and shear modulus (*G*) are related (Figure 5a, Equation (1)), as are Young's modulus and bulk modulus (*K*, Figure 5b, Equation (2)) [278]:

$$G = \frac{E}{2(1 + \upsilon)} \tag{1}$$

$$K = \frac{E}{3(1 - 2\upsilon)} \tag{2}$$

Elasticity theory states that Poisson's ratio must be between −1 and 0.5 for 3D isotropic materials [118,279,280], and between −1 and +1 for 2D isotropic materials [281]. From Equations (1) and (2), as Poisson's ratio tends towards −1 both shear (Equation (1)) and bulk (Equation (2)) modulus are driven towards extremely high or low values (respectively) in a 3D isotropic material.

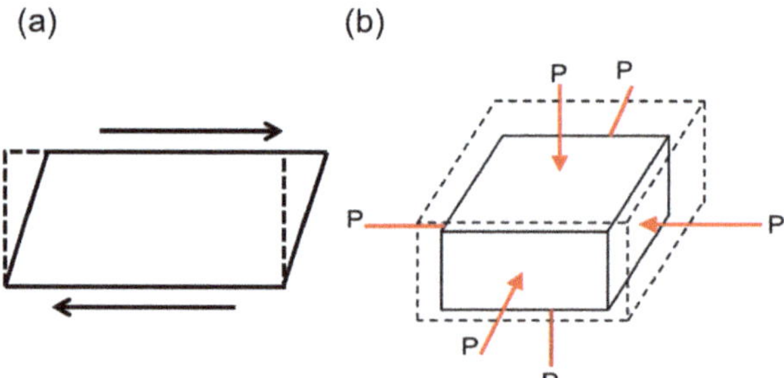

Figure 5. Schematic showing (**a**) shear deformation and (**b**) bulk/volumetric deformation. Black arrows show shear loading, red arrows show loading due to hydrostatic pressure, dashed lines show original (**a**) shape and (**b**) volume.

Indentation resistance is a measure of the load required to indent a material (Figure 6a,b). From elasticity theory, Hertzian indentation resistance (*H*, Equation (3)) for an isotropic material compressed with a uniform indenter depends on Poisson's ratio, Young's modulus and a constant (*x*) related to the shape of the indenter [31].

$$H \propto \left(\frac{E}{(1 - v^2)} \right)^x \tag{3}$$

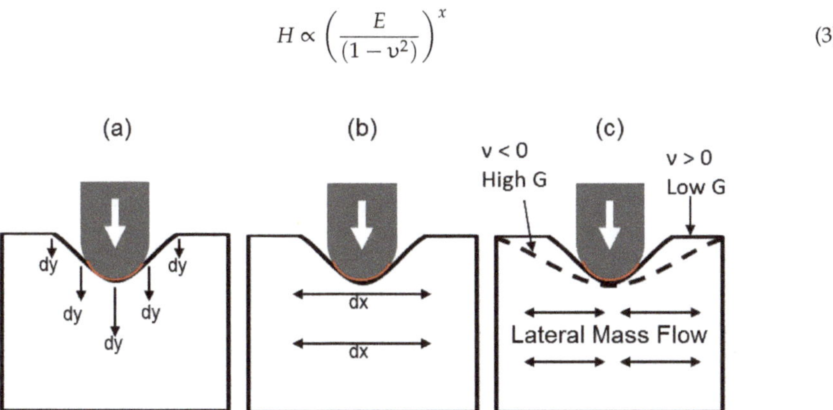

Figure 6. Simplified 2D indentation showing (**a**) Axial deformation; (**b**) Lateral deformation described in Equation (3); (**c**) Effect of large shear modulus (Equation (1)) on indentation area (- - -) and lateral mass flow, both un-accounted for in Equation (3). The red line shows contact area (Equation (4)).

A material with large indentation resistance, but low Young's modulus, could be used in a protective pad that will deform by a similar amount for high or low area (i.e., flat surface vs. studded) impacts. Such a pad would be more versatile, reacting to different surfaces to optimise its resistance to deformation [282]. Assuming a constant Young's modulus, as Poisson's ratio tends towards −1, a material should exhibit infinitely higher resistance to shear deformation (Figure 5a, Equation (1)) and high indentation resistance (Figure 6, Equation (3)), while becoming increasingly easier to deform volumetrically (Figure 5b, Equation (2)). Infinite shear modulus and maximum indentation resistance and volumetric compressibility can only be achieved elastically and isotropically with auxetic materials [10], and increased indentation resistance has been shown experimentally [30,31,283].

Equation (3) for Hertzian indentation resistance comes from elastic properties and assumes; (i) the surfaces are continuous and have non-conforming profiles; (ii) the area of contact (Figure 6) is much smaller than the characteristic dimensions of the contacting bodies; (iii) the strains are small and purely elastic and (iv) the surfaces are frictionless at the contact interface. These four assumptions for Hertzian indentation are not always held, although Equation (3) is often referred to and discussed in the context of non-Hertzian indentations [284,285]. Sporting PPE typically has a low thickness, so does not often meet assumption (ii). A finite thickness model has been developed for soft and thin cushion materials where Hertzian theory is expected to become invalid, and auxetic cushions were found to reduce the contact pressure on the buttocks (indenter) [286]. In another approach, Equation (3) has been adapted for thin sheets of rubber [287], to include a correctional multiplier based on a ratio of contact area between the sheet and indenter (A_t, Figure 6) and the sheet's thickness (a_t) (Equation (4)). As thickness decreases towards zero, contact area/thickness increases and the correction tends towards unity. The force required to indent rubber to a specific depth increases as thickness decreases, but it is unclear if the same trend applies to NPR materials. The final assumption of zero friction has been shown through FEM and continuum mechanics to be invalid in simulations of infield situations [285,288–290]. In these simulations, friction was found to enhance NPR's contribution to indentation resistance.

$$Correction = 1 - exp^{-\frac{A_t}{a_t}} \tag{4}$$

Equations (3) and (4) are for instantaneous, linear values. They do not, therefore, account for the different amounts of densification and possible hardening caused by lateral deformation due to Poisson's ratio (Figure 6c). Auxetic foam (with a relatively high shear modulus, Equation (1)) should resist shear deformation more than its conventional counterpart. Auxetic foam's upper surface would, therefore, compress as a larger, flatter area (represented by the dashed line, Figure 6c) as shown in cylindrical indentations [30] and using FEM [291]. FEM and analytical models also report a reduction in contact area for simulations with NPR materials [292], suggesting that deformation occurred over a larger radius, rather than the foam wrapping around the indenter. The opposite effect (auxetic foam wrapping around the indenter) has been predicted in FEM simulations of low strain indentation of 2D linear-elastic isotropic blocks [183] and also observed during impacts with a hemispherical dropper onto samples covered with 1 to 2 mm thick polypropylene sheets [3]. In the FEM study, a lower Young's modulus as well as NPR was employed for the auxetic foam, providing a shear modulus almost a factor of 2 lower than the conventional foam (Equation (1)), which possibly explains the discrepancy. In the experimental study, the added complexity when considering multi-material systems (e.g., featuring a stiff shell and compliant foam), hemispherical or studded indenters (rather than cylindrical) and high strain rates caused by impact could explain the observed differences. The relationship between the shell and foam's elastic moduli (as discussed in relation to coatings [293,294]), the foam's Poisson's ratio [292,294] and synclastic/anti-synclastic (domed or saddled) bending in the upper surface may affect indentation resistance.

Hertzian indentation requires corrections for high strain indentations, impacts of non-linear materials or multi-material systems typical in sporting PPE [47,48] and, therefore, testing of auxetic foams for sports applications [285,288]. As noted above, friction (assumption iv) is expected to amplify increases in indentation resistance caused by NPR [285,288], and a correction factor has been applied [288]. Experimental indentations of auxetic materials, particularly foams, are not common [7,30], although some limitations of Hertzian indentation theory have been noticed and discussed [30]. Further significant modelling and experimental research examining NPR's effect on indentation responses of conforming, non-linear, anisotropic materials subject to a range of indenter sizes and shapes is therefore required.

One of the difficulties in testing NPR's effect on expected benefits (i.e., impact force attenuation or indentation resistance) when using foam is changes to Young's modulus and stress-strain relationships following fabrication [9,295]. Studies report auxetic foams with lower initial Young's modulus than their conventional counterparts [189,274,295]. Reduced Young's modulus has been attributed to the presence of buckled ribs in the auxetic foam being easier to deform than the straighter ribs of the conventional foam [295]. Elasticity theory (Equation (1), Figure 5a) also supports a reduction in Young's modulus as Poisson's ratio decreases if shear modulus remains constant. Auxetic foams typically have a higher density than their open cell parent foam, so the reduction in Young's modulus is contrary to the usual expectation of an increase in Young's modulus with increased density [118]. Note, though, that Gibson and Ashby refer to a density increase caused by thicker ribs, whereas in the auxetic foam fabrications density increases due to changes in rib orientation. Gibson and Ashby's cellular solid theory actually indicates that Young's modulus can either increase or decrease when moving from a hexagonal to a re-entrant cell geometry, characteristic of auxetic foams [118]. The increased Young's modulus is allowed by elasticity theory—materials with the same bulk modulus (Equation (2), Figure 5b) will have increased Young's modulus as Poisson's ratio decreases. Increases [2,190,197,204] as well as the aforementioned decreases in Young's modulus [9,274,295] have been reported in auxetic foam fabrications.

The re-entrant structures in auxetic foams typically give an initially quasi-linear compressive stress-strain curve, with hardening as pores close at higher compression levels (>~50%) [9,183,185,275,282]. Conventional open cell foams exhibit a low-stiffness plateau region due to buckling of cell ribs between ~5% and 80% strain (Figure 7a) [118]. Both of these cases have been explained numerically and validated experimentally [23,118]. It should be noted that the relatively linear and plateauing stress-strain relationships only apply to specific forms of cellular materials that adapt a re-entrant structure, including PU foams [9,23] and 2D honeycombs [96]. Exceptions have been recently presented; re-entrant auxetic PU foams with a plateau region [23,195].

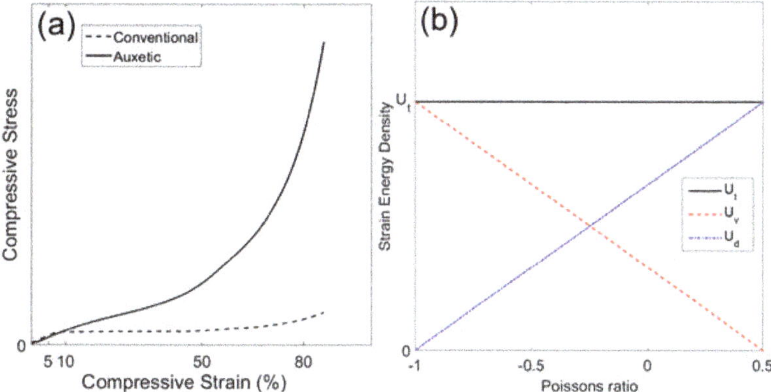

Figure 7. (**a**) Typical compressive stress vs. strain response of auxetic and conventional open cell foam (note cellular collapse between 5 to 10% then hardening beyond 80% strain via densification in open cell foam); (**b**) Strain energy density (Volumetric (U_v, Equation (5)), Distortion (U_d, Equation (6)) & U_t, Total) vs. Poisson's ratio for a linear elastic isotropic material subject to a uniaxial stress. Uniaxial stress and Young's modulus set to 1 kPa.

For a linear elastic isotropic material subject to a uniaxial stress (σ), the total strain energy density (U_t) is the sum of the volumetric strain energy density (U_v) and the distortional strain energy density (U_d), which are related to the Young's modulus and Poisson's ratio (i.e., [278]):

$$U_v = \frac{(1 - 2v)\sigma^2}{6E} \tag{5}$$

$$U_d = \frac{(1 + v)\sigma^2}{3E} \tag{6}$$

In Figure 7b, an applied compressive uniaxial stress and Young's modulus are arbitrarily equated to 1 kPa for simplicity. Plotting these values in Equations (5) and (6) between elastic limits of -1 and 0.5 for Poisson's ratio (Figure 7b) shows an increase in volumetric strain energy density and a reduction in distortional strain energy density as Poisson's ratio tends towards -1. Auxetics material's tendency to volumetric (rather than distortional deformation) could effectively increase indentation resistance (Figure 6c, Equations (3) and (4)). As Poisson's ratio decreases, so does the stress concentration at a crack's tip, preventing crack propagation and increasing toughness [296,297]. Von Mises and maximum shear stress theory both define failure when distortional strain energy exceeds a maximum value. The reduction in distortional strain energy (to zero, Figure 7b) as Poisson's ratio reduces to -1 [279] is, then, expected to lead to an increase in toughness. A natural example of where this may be exploited may be found in the nacre layer of certain seashells. Nacre has a reported tensile NPR of the order of ~-0.1 [93] and ~-0.4 [298] which is thought to increase volumetric strain energy density by ~eleven times while more than halving distortional strain energy density, allowing the system to absorb more energy before failure [298].

Increased energy absorption has been shown experimentally for auxetic foam; under flat plate [33] and studded impacts [274], quasi-statically with flat compression plates [8], within aluminium tubes [299] and with a studded indenter and a stiff shell [7]. When compressed cyclically at high strain rates (0.036 to 0.36 s^{-1} [300] and 0.033 s^{-1} [2,274]) auxetic foams absorbed up to sixteen times more energy than open cell foams of a different polymeric composition and equivalent density, therefore appearing more useful in cushioning layers of sporting (and general) PE & PPE. The differences in energy absorption do not account for changes in stress/strain relationship or strain rate dependency between conventional and auxetic samples. As strain rates increased, Young's modulus and the magnitude of NPR increased marginally in samples of auxetic PU foam [301]. In some cases the increased linear elastic range of the auxetic foam compared to the parent foam with a plateau region in the stress-strain relationship could have contributed to higher energy absorption [2,8,33].

Energy absorption, strain rate dependency and (often) indentation resistance combine to influence performance under impact. Theoretically beneficial for impact protection [282], auxetic foam samples have been shown to exhibit between ~three and ~eight times lower peak force under 2 to 15 J impacts adapted from BS 6183-3:2000 for cricket thigh pads [3,8,55]. During a comparison of high strain rate compression (20 to 40 J) to a conventional commercial foam of similar density, auxetic samples exhibited (1.2 to 1.8 times) higher peak acceleration, but also exhibited higher compressive elastic modulus [2]. Peak forces can be further decreased (1.2 to 1.5 times) during 5 J impacts by impregnating auxetic foam with shear thickening fluid [302].

Inward material flow has been shown under impact by rudimental visual inspection of high-speed camera stills, and the samples with the greatest magnitude of NPR exhibited higher lateral contraction, lower through thickness deformation and a similar peak force to other samples [33]. Newly developed 3D auxetic textile composites exhibit lower peak forces than their conventional counterpart under 12 to 25 J impacts [5]. In helical yarns, a wrap angle of 27° gave the best combination of Poisson's ratio and stiffness for energy absorption [209] during 7 to 65 J impacts.

Honeycomb sandwich panels with auxetic cores were found to resist ballistic impacts better than regular or rectangular cores [303], and absorb more impact energy when the re-entrant cell structure was non-uniform [304]. Laminated composites containing warp knit auxetic Kevlar® fabric reinforcement, under 167 m/s impact with a 14.9 g bullet (~200 J), showed similar energy absorption to laminates containing conventional woven Kevlar® reinforcement [305,306]. The auxetic Kevlar® laminates, however, displayed enhancements in fracture toughness (225%) and fracture initiation toughness (577%), and a reduction in front and rear face damage area [305,306]. Auxetic composite laminates displayed reduced back face damage during 7–18 J impacts by a stud [32,307,308]. Auxetic composite laminates could, therefore, improve the durability of protective shells in PPE and other sports equipment (i.e., bicycle frames or boat hulls).

Related to energy absorption, auxetic foams have been tested for vibration damping (to ISO 13753 for vibration protecting gloves [309]). At low frequencies (<10 Hz), auxetic foam exhibited lower transmissibility than iso-volume open cell samples made from the parent foam and iso-density uni-axially compressed samples [193]. Auxetic foam had a lower cut off frequency than its parent foam [310]. The transmissibility of auxetic samples was greater than 1 between 10 and 31.5 Hz, but less than 1 over 31.5 Hz [183], equivalent to commercial anti-vibration gloves. Auxetic foams also fatigued uniquely, with higher permanent compression than their parent foam and a general increase (as opposed to the conventional foam's decrease) in measured hardness after 80,000 cycles up to ~120 N (150 mm sided cubic samples) [311].

Curvature of a beam or plate subject to an out-of-plane moment is related to Poisson's ratio [9,312]. Sheets with a positive Poisson's ratio will adopt a saddled shape (anticlastic curvature, Figure 8a) and those with NPR will dome (synclastic curvature, Figure 8b). Doming is caused by axial (due to loading) and lateral (due to Poisson's ratio) extension on the upper surface combined with equivalent contractions on the lower surface. Conventional materials will contract laterally on the upper surface and expand laterally on the lower surface.

Figure 8. (a) Conventional honeycomb showing saddled curvature; (b) Re-entrant auxetic honeycomb showing domed curvature. Solid arrows show applied bending and dashed arrows show bending due to Poisson's ratio. Enlarged pop outs show cell structure.

Synclastic curvature has been observed in auxetic foam [9], analysed in detail in sandwich structures with an auxetic honeycomb core and auxetic laminate skins [312] and demonstrated by FEM [37,313] and experimentally using a simple paper model [37]. Auxetic fabric can conform around a spherical surface [36]. Deformation shape could be complicated to predict in nonlinear and often inhomogeneous auxetic foams [4]. Bespoke complex curvatures are achievable for gradient honeycombs displaying conventional and NPR regions when subject to an out-of-plane moment [314].

Some auxetic PU foam samples have been shown to exhibit shape memory, meaning they return to their original dimensions when heated [315] or exposed to solvents [196]. Shape memory auxetic foams investigated by heating in an oven returned rapidly towards their original dimensions when the temperature reached 90 °C. Samples had reached their original dimensions by the time the oven reached the original fabrication temperature of 135 °C.

Marginally re-entrant structures with NPR which exhibit partially blocked shape memory have been fabricated [195]. The fabrication process included numerous cycles of thermo-mechanical fabrication followed by reheating to return samples towards their original state. Auxetic behaviour was found in 'returned' samples from the third returned stage onwards, and has been attributed to the presence of kinked or corrugated ribs. Clearly shape memory could be detrimental in terms of sport safety equipment, as pads could be changed irreparably when exposed to heat or solvents (i.e., when machine washed or dried). Blocking shape memory [195] or investigating solutions to prevent a return to original dimensions (such as constraining auxetic foam in an outer textile/shell layer) could improve a product's lifecycle.

Several characteristics change in the auxetic foam fabrication process, including Poisson's ratio, stress/strain relationships and density [3,4,8,9,23,185,190,316]. So, during comparative impact tests between auxetic and parent foam [2–4,7,8,33,204,316], the specific contribution of individual characteristics, including Poisson's ratio, can be difficult to determine, and unambiguous experimental verification of theoretical enhancements due to the NPR requires further work. The same can be said for other studies into auxetic foam, including those into vibration damping [183,193], resilience/strength [317] and energy absorption [37].

Comparing results from scientific literature suggests that the compressive Young's modulus (30 to 50 kPa) of auxetic open-cell PU foams [3,4,118,185,316] is typically more than twenty times lower than that of the closed cell foams often found in sporting PPE (~1 MPa) [43,47–49,52]. Such a large reduction in stiffness suggests that the two materials are not comparable and stiffer auxetic foam is required for sporting PPE. The enhancements provided by NPR (e.g., indentation resistance) might allow for some reduction in stiffness, but to absorb an equivalent amount of energy to current sporting PPE, current auxetic foam would need a contribution from having NPR that would increase energy absorption by ~twenty times. The largest reported increase in energy absorption for auxetic vs. conventional foam is sixteen times during dynamic cyclic tests [300], but increases of ~three times are more common in single impacts/compressions [8,33,204]. Crash mats are typically softer than PPE (~50 kPa [59]) and (due to their extensive variety of possible applications and impact scenarios) could benefit from the increased energy absorption and indentation resistance associated with auxetic foams [7]. The authors are not aware of any publications specifically comparing impacts or indentations of auxetic foam to foam typically found in sports PE and PPE.

5. The Potential for Auxetic Materials in Sports Products

The sporting goods sector is characterised by early uptake of new technologies and rapid product development, launch and replace cycles. Consequently, this sector is amongst the first to see commercial products based on auxetic materials, with two commercial sports shoe ranges that utilise auxetic structures. The Under Armour Architech sports shoe range [28] incorporates either an AM or moulded auxetic re-entrant latticed upper (Figure 3a) which is claimed to aid conformability around domed shapes, fit and comfort. The added manufacturing benefit of being formable as a one-piece upper, rather than several pieces each individually cut to shape and stitched together is also claimed. The Nike Free RN Flyknit sports shoe [34], on the other hand, employs an architectured closed cell foam outsole with an auxetic rotating triangles structure (shown in Figure 9, first proposed in [105–107]). The outsole is claimed to exhibit bi-axial growth as the wearer accelerates or changes direction, for improved traction and impact energy absorption.

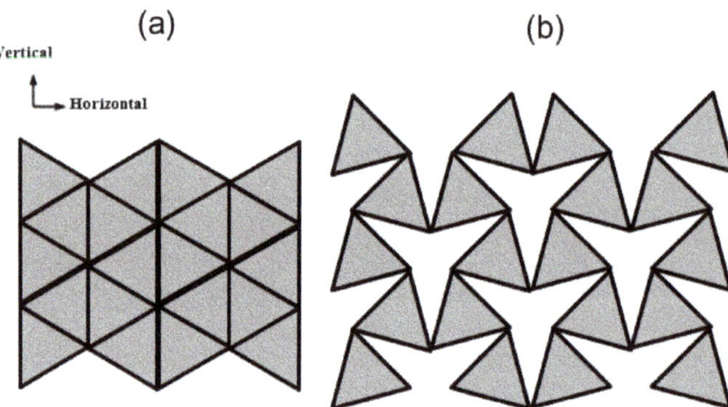

Figure 9. Schematic showing (**a**) rotating triangles (un-deformed); (**b**) rotating triangles' multi axial expansions (rotation = 15°), tensile load applied either vertically or horizontally.

Multi-axis expansion (due to NPR) has potential benefits in cleaning/shedding of dirt [39]. Multi-axis expansion could be implemented in a conformable facemask [35] to protect against extreme weather during outdoor/adventure sports, or commuter cycling—expanding bi-axially when the mouth is opened to increase breathability [9,38–40]. Other benefits of multi-axis expansion include large volume change, allowing design of small, easily stored garments which can expand when worn (i.e., for spare layers in outdoor/adventure sports).

D3O market the Trust Helmet Pad System [318] incorporating pads with a re-entrant auxetic geometry that are claimed to provide increased fit to the head and decreased acceleration under blunt impact. More products are likely to emerge if the increased level of comfort and protection offered by auxetic materials can be further demonstrated, justifying investment in the development of stiffer, more appropriate auxetic materials (such as closed cell [199] or shear thickening fluid impregnated [302] auxetic foams, or novel impact hardening auxetic polymers [319]). The emergence of sports products including auxetic materials could be assisted by the development of a commercially-viable (low cost, large scale) auxetic foam production process [23,185,186]. 'Felted' foams, fabricated by uniaxial compression, are already commercially produced and exhibit auxetic behaviour in one loading direction [33]). Alternatively AM is a commercially viable option [28] to provide fine control over unit cell structure for auxetic foam-like structures.

The ability to change the shear modulus of a material through Poisson's ratio [118,151] could provide a useful tool in designing head protection that can reduce rotational acceleration [6]. When used in place of a comfort layer in a helmet, auxetic foam can significantly reduce the severity of direct impacts in comparison to its conventional counterpart [320]. Helmets and helmet certification standards have been criticised for focussing solely on direct rather than both direct and oblique impacts [69]. Controlling shear modulus by changing Poisson's ratio (to negative or positive values) of any layer within the helmet could contribute to solutions to reduce rotational acceleration (i.e., in combination with or instead of slip plane technology [77]). The benefit of using Poisson's ratio to change shear modulus is that elastic moduli would change less, limiting the effect on linear acceleration and therefore perceived performance according to current certification standards [66,67].

6. Conclusions

In a competitive, rapid uptake market such as sports equipment, it is important to keep searching for new and improved designs and materials. With frequent, catastrophic high-profile injuries and scandals (particularly involving head injury) the importance of improvement increases. The chance to increase market revenue, while reducing costs to health services, national economies and burdens on injured individuals, warrants continued research through commercial and state investment.

Auxetics have a wide range of potentially useful characteristics, including increased indentation resistance, vibration damping and shear modulus as well as decreased bulk modulus. These have been shown using auxetic foam compared to its parent foam and iso-density conventional foams. Further experimental comparisons of auxetic and conventional foam's indentation resistance, impact force attenuation and vibration damping are required to determine contributions from other variables, such as elastic modulus and density, to clearly identify the contribution due to Poisson's ratio.

The creation of auxetic foams appropriate and beneficial to sporting goods will require development of fabrication processes, especially for larger samples. Fabricating samples with comparative characteristics (e.g., Young's modulus and density) is also required to show the NPR's benefits (i.e., to indentation resistance). The solvent and CO_2 softening routes provide alternative and, in the latter case, faster auxetic foam fabrication. Increasing the stiffness of auxetic foam closer to the closed cell foam found in sporting PPE or ensuring that impact energy absorption is equivalent or higher is an important step for commercialisation. With dynamic energy absorption reported up to sixteen times greater in auxetic foams some reduction in stiffness is likely acceptable, but requires confirmation by appropriate testing. AM presents an alternative for the creation of auxetic cellular solids with bespoke properties. There is increasing confidence, then, that commercial auxetic foam production will be achievable, either by AM or by improved fabrication methods from open cell foams. Commercially available auxetic foam would improve access, allowing more sport safety equipment manufacturers to design, test and evaluate auxetic prototypes.

Another option to increase auxetic materials range of moduli and impact force attenuation is combining materials. Auxetic foam's impact force attenuation increased up to ~1.5 times when supported by shear thickening fluid. Enhancing the characteristics of auxetic fabrics and textiles by combining with a conventional, gradient or auxetic material could facilitate the production of sports garments and performance apparel that deforms with the wearer and attenuates impact forces. Combining conventional and auxetic sections of honeycombs in gradient sandwich structures increases bending stiffness close to transitions. Snowboards, skis, tennis rackets and hockey sticks (to name a few) use conventional sandwich structures, and could benefit from increased rigidity or equivalent rigidity and lower mass.

The more obvious characteristics of auxetic materials that are backed up with strong supporting evidence (multi axis expansion and domed curvature) have been implemented as auxetic foam and AM materials into sporting footwear and helmet pad products. Commercial success for auxetic sporting PPE requires stronger supporting evidence that responds to trends accurately re-creating infield collisions and falls. Other auxetic materials such as laminates and knitted/woven fabrics that exhibit characteristics including increased fracture toughness and, potentially, tailorable shape change, are currently untested for sports applications within the peer-reviewed scientific literature. Unlike auxetic foam production, the fabrication of auxetic composite laminates and many of the auxetic fabrics is via established commercial processes, requiring little or no modification. With further development and testing auxetic laminates and fabrics could be applied to a range of sporting products, from carbon fibre reinforced composite bicycle frames to swim suits or rugby tops that deform with the movements of the wearer.

Conflicts of Interest: The authors declare no conflict of interest.

References

1. Evans, K.E. Auxetic polymers: A new range of materials. *Endeavour* **1991**, *15*, 170–174. [CrossRef]
2. Lisiecki, J.; Błazejewicz, T.; Kłysz, S.; Gmurczyk, G.; Reymer, P.; Mikułowski, G. Tests of polyurethane foams with negative Poisson's ratio. *Phys. Status Solidi Basic Res.* **2013**, *250*, 1988–1995. [CrossRef]
3. Allen, T.; Shepherd, J.; Hewage, T.A.M.; Senior, T.; Foster, L.; Alderson, A. Low-kinetic energy impact response of auxetic and conventional open-cell polyurethane foams. *Phys. Status Solidi Basic Res.* **2015**, *9*, 1631–1639. [CrossRef]
4. Duncan, O.; Foster, L.; Senior, T.; Alderson, A.; Allen, T. Quasi-static characterisation and impact testing of auxetic foam for sports safety applications. *Smart Mater. Struct.* **2016**, *25*, 54014. [CrossRef]
5. Zhou, L.; Zeng, J.; Jiang, L.; Hu, H. Low-velocity impact properties of 3D auxetic textile composite. *J. Mater. Sci.* **2017**, *53*, 3899–3914. [CrossRef]
6. Taha, Z.; Hassan, M.H.A. Parametric Analysis of the Influence of Elastomeric Foam on the Head Response during Soccer Heading Manoeuvre. *Procedia Eng.* **2016**, *147*, 139–144. [CrossRef]
7. Allen, T.; Duncan, O.; Foster, L.; Senior, T.; Zampieri, D.; Edeh, V.; Alderson, A. Auxetic foam for snowsport safety devices. In *Snow Sports Trauma and Safety, Proceedings of the International Society of Skiing Safety*; Springer: Cham, Switzerland, 2016; pp. 145–159.
8. Duncan, O.; Foster, L.; Senior, T.; Allen, T.; Alderson, A. A Comparison of Novel and Conventional Fabrication Methods for Auxetic Foams for Sports Safety Applications. *Procedia Eng.* **2016**, *147*, 384–389. [CrossRef]
9. Lakes, R.S. Foam Structures with a Negative Poisson's Ratio. *Science* **1987**, *235*, 1038–1041. [CrossRef] [PubMed]
10. Evans, K.E.; Alderson, A. Auxetic materials: Functional materials and structures from lateral thinking! *Adv. Mater.* **2000**, *12*, 617–628. [CrossRef]
11. Yang, W.; Li, Z.M.; Shi, W.; Xie, B.H.; Yang, M.B. On auxetic materials. *J. Mater. Sci.* **2004**, *39*, 3269–3279. [CrossRef]
12. Greaves, G.N.; Greer, A.L.; Lakes, R.S.; Rouxel, T. Poisson's ratio and modern materials. *Nat. Mater.* **2011**, *10*, 823–837. [CrossRef] [PubMed]
13. Critchley, R.; Corni, I.; Wharton, J.A.; Walsh, F.C.; Wood, R.J.K.; Stokes, K.R. A review of the manufacture, mechanical properties and potential applications of auxetic foams. *Phys. Status Solidi Basic Res.* **2013**, *250*, 1963–1982. [CrossRef]
14. Prawoto, Y. Seeing auxetic materials from the mechanics point of view: A structural review on the negative Poisson's ratio. *Comput. Mater. Sci.* **2012**, *58*, 140–153. [CrossRef]
15. Kolken, H.M.A.; Zadpoor, A.A. Auxetic mechanical metamaterials. *RSC Adv.* **2017**, *7*, 5111–5129. [CrossRef]
16. Lakes, R.S. Negative-Poisson's-Ratio Materials: Auxetic Solids. *Annu. Rev. Mater. Res.* **2017**, *47*, 63–81. [CrossRef]
17. Ren, X.; Das, R.; Tran, P.; Ngo, T.D.; Xie, Y.M. Auxetic metamaterials and structures: A review. *Smart Mater. Struct.* **2018**, *27*, 23001. [CrossRef]
18. Schmitt, K.U.; Liechti, B.; Michel, F.I.; Stämpfli, R.; Brühwiler, P.A. Are current back protectors suitable to prevent spinal injury in recreational snowboarders? *Br. J. Sports Med.* **2010**, *44*, 822–826. [CrossRef] [PubMed]

19. Mcintosh, A.S. Biomechanical considerations in the design of equipment to prevent sports injury. *Proc. Inst. Mech. Eng. Part P* **2011**, *226*, 193–199. [CrossRef]
20. Li, Y.; Zeng, C. On the successful fabrication of auxetic polyurethane foams: Materials requirement, processing strategy and conversion mechanism. *Polymer* **2016**, *87*, 98–107. [CrossRef]
21. Li, Y.; Zeng, C. Room-Temperature, Near-Instantaneous Fabrication of Auxetic Materials with Constant Poisson's Ratio over Large Deformation. *Adv. Mater.* **2016**, *28*, 2822–2826. [CrossRef] [PubMed]
22. Sanami, M.; Alderson, A.; Alderson, K.L.; McDonald, S.A.; Mottershead, B.; Withers, P.J. The production and characterization of topologically and mechanically gradient open-cell thermoplastic foams. *Smart Mater. Struct.* **2014**, *23*, 55016. [CrossRef]
23. Duncan, O.; Allen, T.; Foster, L.; Senior, T.; Alderson, A. Fabrication, characterisation and modelling of uniform and gradient auxetic foam sheets. *Acta Mater.* **2017**, *126*, 426–437. [CrossRef]
24. Gatt, R.; Attard, D.; Farrugia, P.S.; Azzopardi, K.M.; Mizzi, L.; Brincat, J.P.; Grima, J.N. A realistic generic model for anti-tetrachiral systems. *Phys. Status Solidi Basic Res.* **2013**, *250*, 2012–2019. [CrossRef]
25. Yang, L.; Harrysson, O.; West, H.; Cormier, D. Modeling of uniaxial compression in a 3D periodic re-entrant lattice structure. *J. Mater. Sci.* **2013**, *48*, 1413–1422. [CrossRef]
26. Scarpa, F.; Panayiotou, P.; Tomlinson, G. Numerical and experimental uniaxial loading on in-plane auxetic honeycombs Numerical and experimental uniaxial loading on in-plane auxetic honeycombs. *J. Strain Anal. Eng. Des.* **2000**, *35*, 383–388. [CrossRef]
27. Wang, Y.; Ma, Z.D.; Wang, L. A finite element stratification method for a polyurethane jounce bumper. *Proc. Inst. Mech. Eng. Part D J. Automob. Eng.* **2016**, *230*, 983–992. [CrossRef]
28. Toronjo, A. Articles of Apparel Including Auxetic Materials. U.S. Patent 20140059734 A1, 6 March 2014.
29. Liu, Y.; Hu, H.; Lam, J.K.C.; Liu, S. Negative Poisson's Ratio Weft-knitted Fabrics. *Text. Res. J.* **2010**, *80*, 856–863.
30. Chan, N.; Evans, K.E. Indentation resilience of conventional and auxetic foams. *J. Cell. Plast.* **1998**, *34*, 231–260. [CrossRef]
31. Alderson, K.L.; Pickles, A.P.; Neale, P.J.; Evans, K.E. Auxetic polyethylene: The effect of a negative poisson's ratio on hardness. *Acta Metall. Mater.* **1994**, *42*, 2261–2266. [CrossRef]
32. Alderson, K.L.; Simkins, V.R.; Coenen, V.L.; Davies, P.J.; Alderson, A.; Evans, K.E. How to make auxetic fibre reinforced composites. *Phys. Status Solidi Basic Res.* **2005**, *242*, 509–518. [CrossRef]
33. Ge, C. A comparative study between felted and triaxial compressed polymer foams on cushion performance. *J. Cell. Plast.* **2013**, *49*, 521–533. [CrossRef]
34. Cross, T.M.; Hoffer, K.W.; Jones, D.P.; Kirschner, P.B.; Meschter, J.C. Auxetic Structures and Footwear with Soles Having Auxetic Structures. U.S. Patent 2015/0075034 A1, 19 March 2015.
35. Martin, P.G. Filtering Face-Piece Respirator Having an Auxetic Mesh in the Mask Body. U.S. Patent 2011/0155137 A1, 30 June 2011.
36. Wang, Z.; Hu, H. 3D auxetic warp-knitted spacer fabrics. *Phys. Status Solidi Basic Res.* **2014**, *251*, 281–288. [CrossRef]
37. Sanami, M.; Ravirala, N.; Alderson, K.; Alderson, A. Auxetic materials for sports applications. *Procedia Eng.* **2014**, *72*, 453–458. [CrossRef]
38. Alderson, A.; Evans, K.E.; Rasburn, J. Separation Process and Apparatus. WO Patent 1999022838A1, 14 May 1999.
39. Alderson, A.; Rasburn, J.; Evans, K.E. An Auxetic Filter: A Tuneable Filter Displaying Enhanced Size Selectivity or Defouling Properties. *Ind. Eng. Chem. Res.* **2000**, *39*, 654–665. [CrossRef]
40. Alderson, A.; Rasburn, J.; Evans, K.E. Mass transport properties of auxetic (negative Poisson's ratio) foams. *Phys. Status Solidi* **2007**, *244*, 817–827. [CrossRef]
41. Hou, Y.; Neville, R.; Scarpa, F.; Remillat, C.; Gu, B.; Ruzzene, M. Graded conventional-auxetic Kirigami sandwich structures: Flatwise compression and edgewise loading. *Compos. Part B Eng.* **2014**, *59*, 33–42. [CrossRef]
42. Hou, Y.; Tai, Y.H.; Lira, C.; Scarpa, F.; Yates, J.R.; Gu, B. The bending and failure of sandwich structures with auxetic gradient cellular cores. *Compos. Part A Appl. Sci. Manuf.* **2013**, *49*, 119–131. [CrossRef]
43. Hrysomallis, C.; Morrison, W.; He, J. Assessing the shock absorption of thigh pads. *J. Sci. Med.* **1999**, *2*, 49. [CrossRef]

44. Payne, T.; Mitchell, S.; Halkon, B.; Bibb, R. A systematic approach to the characterisation of human impact injury scenarios in sport. *BMJ Open Sport Exerc. Med.* **2016**, *2*, e000017. [CrossRef] [PubMed]

45. Schmikli, S.L.; Backx, F.J.G.; Kemler, H.J.; Van Mechelen, W. National survey on sports injuries in the netherlands: Target populations for sports injury prevention programs. *Clin. J. Sport Med.* **2009**, *19*, 101–106. [CrossRef] [PubMed]

46. Statista. Sporting Goods Industry—Statistics & Facts [Internet]. 2018. Available online: https://www.statista.com/topics/961/sporting-goods/ (accessed on 1 May 2018).

47. Ankrah, S.; Mills, N.J. Analysis of ankle protection in Association football. *Sport Eng.* **2004**, *7*, 41–52. [CrossRef]

48. Ankrah, S.; Mills, N.J. Performance of football shin guards for direct stud impacts. *Sport Eng.* **2003**, *6*, 207–219. [CrossRef]

49. Hrysomallis, C. Surrogate thigh model for assessing impact force attenuation of protective pads. *J. Sci. Med. Sport* **2009**, *12*, 35–41. [CrossRef] [PubMed]

50. Michel, F.I.; Schmitt, K.U.; Liechti, B.; Stämpfli, R.; Brühwiler, P. Functionality of back protectors in snow sports concerning safety requirements. *Procedia Eng.* **2010**, *2*, 2869–2874. [CrossRef]

51. Adams, C.; James, D.; Senior, T.; Allen, T.; Hamilton, N. Development of a Method for Measuring Quasi-static Stiffness of Snowboard Wrist Protectors. *Procedia Eng.* **2016**, *147*, 378–383. [CrossRef]

52. Mills, N.J. The biomechanics of hip protectors. *Proc. Inst. Mech. Eng. Part H* **1996**, *210*, 259–266. [CrossRef] [PubMed]

53. Jenkins, M. *Materials in Sports Equipment*; Woodhead Publishing Ltd.: Cambridge, UK, 2007; Volume 2, pp. 138–139.

54. American Society for Testing and Materials. *ASTM F2040-02: Standard Specification for Helmets Used for Recreational Snow Sports*; ASTM International: West Conshohocken, PA, USA, 2002; Volume 15, pp. 1–4.

55. British Standards Institution. *BS 6183-3:2000-Protective Equipment for Cricketers*; British Standards Institution: London, UK, 2000.

56. British Standards Institution. *Helmets for Alpine Skiers and Snowboarders*; BS EN 1077:2007; British Standards Institution: London, UK, 2007; Volume 3.

57. European Committee for Standardization. *Protective Clothing—Shin Guards for Association Football Players—Requirements and Test Methods*; BS EN: 13061:2009; European Committee for Standardization: Brussels, Belgium, 2009.

58. World Fightsport & Martial Arts Council. Official Rulebook. 2010. Available online: www.wfmc-kickboxing.com (accessed on 29 June 2017).

59. Lyn, G.; Mills, N.J. Design of foam crash mats for head impact protection. *Sport. Eng.* **2001**, *4*, 153–163. [CrossRef]

60. British Standards Institution. *Sports Mats, Part 1: Gymnastic Mats, Safety Requirements*; British Standards Institution: London, UK, 2013.

61. European Committee for Standardization. *Protective Clothing—Wrist, Palm, Knee and Elbow Protectors for Users of Roller Sports Equipment—Requirements and Test Methods*; EN 14120:2003; European Committee for Standardization: Brussels, Belgium, 2003.

62. European Committee for Standardization. *Motorradfahrer Schutzkleidung Teil 2 Rückenprotektoren*; EN 1621-2: 2003; European Committee for Standardization: Brussels, Belgium, 2003.

63. Dickson, T.J.; Trathen, S.; Terwiel, F.A.; Waddington, G.; Adams, R. Head injury trends and helmet use in skiers and snowboarders in Western Canada, 2008–2009 to 2012–2013: An ecological study. *Scand. J. Med. Sci. Sports* **2017**, *27*, 236–244. [CrossRef] [PubMed]

64. Ekeland, A.; Rødven, A.; Heir, S. Injury Trends in Recreational Skiers and Boarders in the 16-Year Period 1996–2012. In *Snow Sports Trauma and Safety*; Scher, I.S., Greenwald, R.M., Petrone, N., Eds.; Springer: Cham, Switzerland, 2017; pp. 3–16.

65. Mez, J.; Daneshvar, D.H.; Kiernan, P.T.; Abdolmohammadi, B.; Alvarez, V.E.; Huber, B.R.; Alosco, M.L.; Solomon, T.M.; Nowinski, C.J.; McHale, L.; et al. Clinicopathological Evaluation of Chronic Traumatic Encephalopathy in Players of American Football. *JAMA* **2017**, *318*, 360–370. [PubMed]

66. King, A.I.; Yang, K.H.; Zhang, L.; Hardy, W.; Viano, D. Is Head Injury Caused by Linear or Angular Acceleration? In Proceedings of the IRCOBI Conference, Lisbon, Portugal, 25–26 September 2003; pp. 1–12.

67. Rowson, S.; Duma, S.M. Brain injury prediction: Assessing the combined probability of concussion using linear and rotational head acceleration. *Ann. Biomed. Eng.* **2013**, *41*, 873–882. [CrossRef] [PubMed]

68. National Operating Committee on Standards for Athletic Equipment (NOCSAE). *Standard Test Method and Equipment Used in Evaluating the Performance Characteristics of Protective Headgear/Equipment*; DOC. 001-13m15; NOCSAE: Overland Park, KS, USA, 2011.

69. Mcintosh, A.S.; Andersen, T.E.; Bahr, R.; Greenwald, R.; Kleiven, S.; Turner, M.; Varese, M.; McCrory, P. Sports helmets now and in the future. *Br. J. Sports Med.* **2011**, *45*, 1258–1265. [CrossRef] [PubMed]

70. Casson, I.R.; Viano, D.C.; Powell, J.W.; Pellman, E.J. Twelve years of National Football League concussion data. *Sports Health* **2010**, *2*, 471–483. [CrossRef] [PubMed]

71. British Standards Institution. *Protective Helmets for Vehicle Users*; BS 6658:1985; British Standards Institution: London, UK, 1985.

72. British Standards Institution. *Specification for Head Protectors for Cricketers*; BS 7928:2013; British Standards Institution: London, UK, 2013.

73. McIntosh, A.S.; Janda, D. Evaluation of cricket helmet performance and comparison with baseball and ice hockey helmets. *Br. J. Sports Med.* **2003**, *37*, 325–330. [CrossRef] [PubMed]

74. Van Bekkum, J.E.; Williams, J.M.; Morris, P.G. Cycle commuting and perceptions of barriers: Stages of change, gender and occupation. *Health Educ.* **2011**, *111*, 476–497. [CrossRef]

75. Heinen, E.; Maat, K.; Van Wee, B. The role of attitudes toward characteristics of bicycle commuting on the choice to cycle to work over various distances. *Transp. Res. Part D* **2011**, *16*, 102–109. [CrossRef]

76. Willinger, R.; Deck, C.; Halldin, P.; Otte, D. Towards advanced bicycle helmet test methods. In Proceedings of the International Cycling Safety Conference, Göteborg, Sweden, 18–19 November 2014; pp. 1–11.

77. Zuzarte, P. Protective Helmet. U.S. Patent 6,658,671 B1, 9 December 2003.

78. Aare, M.; Kleiven, S.; Halldin, P. Injury tolerances for oblique impact helmet testing. *Int. J. Crashworthiness* **2004**, *9*, 15–23. [CrossRef]

79. Kleiven, S. Influence of impact direction on the human head in prediction of subdural hematoma. *J. Neurotrauma* **2003**, *20*, 365–379. [CrossRef] [PubMed]

80. Kleiven, S. Evaluation of head injury criteria using a finite element model validated against experiments on localized brain motion, intracerebral acceleration, and intracranial pressure. *Int. J. Crashworthiness* **2006**, *11*, 65–79. [CrossRef]

81. Kleiven, S.; Hardy, W.N. Correlation of an FE Model of the Human Head with Local Brain Motion—Consequences for Injury Prediction. *Stapp Car Crash J.* **2002**, *46*, 123–144. [PubMed]

82. Dura, J.V.; Garcia, A.C.; Solaz, J. Testing shock absorbing materials: The application of viscoelastic linear model. *Sport Eng.* **2002**, *5*, 9–14. [CrossRef]

83. Hayes, S.G.; Venkatraman, P. *Materials and Technology for Sportswear and Performance Apparel*; CRC Press: Boca Raton, FL, USA, 2016; p. 314.

84. Adams, C.; James, D.; Senior, T.; Allen, T.; Hamilton, N. Effect of surrogate design on the measured stiffness of snowboarding wrist protectors. *Sport Eng.* **2018**, *1*, 1–9.

85. Michel, F.I.; Schmitt, K.U.; Greenwald, R.M.; Russell, K.; Simpson, F.I.; Schulz, D.; Langran, M. White Paper: Functionality and efficacy of wrist protectors in snowboarding-towards a harmonized international standard. *Sport Eng.* **2013**, *16*, 197–210. [CrossRef]

86. Payne, T.; Mitchell, S.; Halkon, B.; Bibb, R.; Waters, M. Development of a synthetic human thigh impact surrogate for sports personal protective equipment testing. *Proc. Inst. Mech. Eng. Part P* **2016**, *230*, 5–16. [CrossRef]

87. Petrone, N.; Carraro, G.; Dal Castello, S.; Broggio, L.; Koptyug, A.; Backstrom, M. A novel instrumented human head surrogate for the impact evaluation of helmets. *Proceedings* **2018**, *2*, 269. [CrossRef]

88. Nakamura, K.E.N.I.; Wada, M.; Kuga, S.; Okano, T. Poisson's Ratio of Cellulose I_β and Cellulose II. *J. Polym. Sci. Part B Polym. Phys.* **2004**, *42*, 1206–1211. [CrossRef]

89. Caddock, B.D.; Evans, K.E. Microporous materials with negative Poisson's ratios. I. Microstructure and mechanical properties. *J. Phys. D. Appl. Phys.* **1989**, *22*, 1877–1882. [CrossRef]

90. Dominec, J.; Vase, P.; Svoboda, P.; Plechacek, V.; Laermans, C. Elastic Moduli for Three Superconducting Phases of Bi-Sr-Ca-Cu-O. *Mod. Phys. Lett. B* **1992**, *6*, 1049. [CrossRef]

91. Baughman, R.H.; Shacklette, J.M.; Zakhidov, A.A.; Stafstro, S. Negative Poisson's ratios as a common feature of Cubic Metals. *Nature* **1998**, *392*, 362–365. [CrossRef]

92. Clarke, J.F.; Duckett, R.A.; Hine, P.J.; Hutchinson, I.J.; Ward, I.M. Negative Poisson's ratios in angle-ply laminates: Theory and experiment. *Composites* **1994**, *25*, 863–868. [CrossRef]

93. Barthelat, F.; Tang, H.; Zavattieri, P.D.; Li, C.M.; Espinosa, H.D. On the mechanics of mother-of-pearl: A key feature in the material hierarchical structure. *J. Mech. Phys. Solids* **2007**, *55*, 306–337. [CrossRef]

94. Rad, M.S.; Mohsenizadeh, S.; Ahmad, Z. Finite element approach and mathematical formulation of viscoelastic auxetic honeycomb structures for impact mitigation. *J. Eng. Sci. Technol.* **2017**, *12*, 471–490.

95. Gibson, L.J.; Ashby, M.F.; Schajer, G.S.; Robertson, C.I. The mechanics of two-dimensional cellular materials. *Proc. R. Soc. Lond. A* **1982**, *382*, 25–42. [CrossRef]

96. Masters, I.G.; Evans, K.E. Models for the elastic deformation of honeycombs. *Compos. Struct.* **1996**, *35*, 403–422. [CrossRef]

97. Whitty, J.P.M.; Nazare, F.; Alderson, A. Modelling the effects of density variations on the in-plane Poisson's ratios and Young's moduli of periodic conventional and re-entrant honeycombs—Part 1: Rib thickness variations. *Cell. Polym.* **2002**, *21*, 69–98.

98. Nkansah, M.A.; Evans, K.E.; Hutchinson, I.J. Modelling the mechanical properties of an auxetic molecular network. *Model. Simul. Mater. Sci. Eng.* **1994**, *2*, 337–352. [CrossRef]

99. Strek, T.; Maruszewski, B.; Narojczyk, J.W.; Wojciechowski, K.W. Finite element analysis of auxetic plate deformation. *J. Non-Cryst. Solids* **2008**, *354*, 4475–4480. [CrossRef]

100. Ge, Z.; Hu, H.; Liu, Y. A finite element analysis of a 3D auxetic textile structure for composite reinforcement. *Smart Mater. Struct.* **2013**, *22*, 84005. [CrossRef]

101. Zhang, J.; Lu, G.; Wang, Z.; Ruan, D.; Alomarah, A.; Durandet, Y. Large deformation of an auxetic structure in tension: Experiments and finite element analysis. *Compos. Struct.* **2018**, *184*, 92–101. [CrossRef]

102. Almgren, R.F. An isotropic three-dimensional structure with Poisson's ratio = −1. *J. Elast.* **1985**, *15*, 427–430.

103. Wojciechowski, K.W. Constant thermodynamic tension Monte Carlo studies of elastic properties of a two-dimensional system of hard cyclic hexamers. *Mol. Phys.* **1987**, *61*, 1247–1258. [CrossRef]

104. Prall, D.; Lakes, R.S. Properties of A Chiral Honeycomb with A Poisson's Ratio of—1. *Int. J. Mech. Sci.* **1997**, *39*, 305–314. [CrossRef]

105. Grima, J.N.; Evans, K.E. Auxetic behavior from rotating squares. *J. Mater. Sci. Lett.* **2000**, *19*, 1563–1565. [CrossRef]

106. Grima, J.N.; Alderson, A.; Evans, K.E. Auxetic behaviour from rotating rigid units. *Phys. Status Solidi Basic Res.* **2005**, *242*, 561–575. [CrossRef]

107. Grima, J.N.; Farrugia, P.S.; Gatt, R.; Zammit, V. Connected Triangles Exhibiting Negative Poisson's Ratios and Negative Thermal Expansion. *J. Phys. Soc. Jpn.* **2007**, *76*, 025001. [CrossRef]

108. Attard, D.; Grima, J.N. A three-dimensional rotating rigid units network exhibiting negative Poisson's ratios. *Phys. Status Solidi Basic Res.* **2012**, *249*, 1330–1338. [CrossRef]

109. Grima, J.N.; Gatt, R.; Ravirala, N.; Alderson, A.; Evans, K.E. Negative Poisson's ratios in cellular foam materials. *Mater. Sci. Eng. A* **2006**, *423*, 214–218. [CrossRef]

110. Smith, C.W.; Grima, J.N.; Evans, K.E. Novel mechanism for generating auxetic behaviour in reticulated foams: Missing rib foam model. *Acta Mater.* **2000**, *48*, 4349–4356. [CrossRef]

111. Bertoldi, K. Harnessing Instabilities to Design Tunable Architected Cellular Materials. *Annu. Rev. Mater. Res.* **2017**, *47*, 51–61. [CrossRef]

112. Alderson, A.; Evans, K.E. Modelling concurrent deformation mechanisms in auxetic microporous polymers. *J. Mater. Sci.* **1997**, *32*, 2797–2809. [CrossRef]

113. Bertoldi, K.; Reis, P.M.; Willshaw, S.; Mullin, T. Negative poisson's ratio behavior induced by an elastic instability. *Adv. Mater.* **2010**, *22*, 361–366. [CrossRef] [PubMed]

114. Overvelde, J.T.B.; Shan, S.; Bertoldi, K. Compaction through buckling in 2D periodic, soft and porous structures: Effect of pore shape. *Adv. Mater.* **2012**, *24*, 2337–2342. [CrossRef] [PubMed]

115. Overvelde, J.T.B.; Bertoldi, K. Relating pore shape to the non-linear response of periodic elastomeric structures. *J. Mech. Phys. Solids* **2014**, *64*, 351–366. [CrossRef]

116. Babaee, S.; Shim, J.; Weaver, J.C.; Chen, E.R.; Patel, N.; Bertoldi, K. 3D soft metamaterials with negative poisson's ratio. *Adv. Mater.* **2013**, *25*, 5044–5049. [CrossRef] [PubMed]

117. Javid, F.; Smith-Roberge, E.; Innes, M.C.; Shanian, A.; Weaver, J.C.; Bertoldi, K. Dimpled elastic sheets: A new class of non-porous negative Poisson's ratio materials. *Sci. Rep.* **2015**, *5*, 1–9. [CrossRef] [PubMed]

118. Gibson, L.J.; Ashby, M.F. *Cellular Solids. Structure and Properties*; Cambridge University Press: Cambridge, UK, 1997; pp. 67, 176–183, 259–264, 286, 498.

119. Choi, J.B.; Lakes, R.S. Analysis of elastic modulus of conventional foams and of re-entrant foam materials with a negative Poisson's ratio. *Int. J. Mech. Sci.* **1995**, *37*, 51–59. [CrossRef]

120. Evans, K.E.; Alderson, A.; Christian, F.R. Auxetic Two-dimensional Polymer Networks. *J. Chem. Soc. Faraday Trans.* **1995**, *91*, 2671–2680. [CrossRef]

121. Alipour, M.M.; Shariyat, M. Analytical zigzag formulation with 3D elasticity corrections for bending and stress analysis of circular/annular composite sandwich plates with auxetic cores. *Compos. Struct.* **2015**, *132*, 175–197. [CrossRef]

122. Wu, W.; Song, X.; Liang, J.; Xia, R.; Qian, G.; Fang, D. Mechanical properties of anti-tetrachiral auxetic stents. *Compos. Struct.* **2018**, *185*, 381–392. [CrossRef]

123. Abramovitch, H.; Burgard, M.; Edery-Azulay, L.; Evans, K.E.; Hoffmeister, M.; Miller, W.; Scarpa, F.; Smith, C.W.; Tee, K.F. Smart tetrachiral and hexachiral honeycomb: Sensing and impact detection. *Compos. Sci. Technol.* **2010**, *70*, 1072–1079. [CrossRef]

124. Grima, J.N.; Attard, D.; Ellul, B.; Gatt, R. An improved analytical model for the elastic constants of auxetic and conventional hexagonal honeycombs. *Cell. Polym.* **2011**, *30*, 287–310.

125. Hughes, T.P.; Marmier, A.; Evans, K.E. Auxetic frameworks inspired by cubic crystals. *Int. J. Solids Struct.* **2010**, *47*, 1469–1476. [CrossRef]

126. Wang, X.T.; Li, X.W.; Ma, L. Interlocking assembled 3D auxetic cellular structures. *Mater. Des.* **2016**, *99*, 467–476. [CrossRef]

127. Zhang, J.; Lu, G.; Ruan, D.; Wang, Z. Tensile behavior of an auxetic structure: Analytical modeling and finite element analysis. *Int. J. Mech. Sci.* **2018**, *136*, 143–154. [CrossRef]

128. Dirrenberger, J.; Forest, S.; Jeulin, D. Elastoplasticity of auxetic materials. *Comput. Mater. Sci.* **2012**, *64*, 57–61. [CrossRef]

129. Najarian, F.; Alipour, R.; Shokri Rad, M.; Nejad, A.F.; Razavykia, A. Multi-objective optimization of converting process of auxetic foam using three different statistical methods. *Meas. J. Int. Meas. Confed.* **2018**, *119*, 108–116. [CrossRef]

130. Evans, K.E.; Nkansah, M.A.; Hutchinson, I.J. Auxetic Foams: Modelling Negative Poisson's Ratios. *Acta Metall. Mater.* **1994**, *42*, 1289–1294. [CrossRef]

131. Crespo, J.; Montáns, F.J. A continuum approach for the large strain finite element analysis of auxetic materials. *Int. J. Mech. Sci.* **2018**, *135*, 441–457. [CrossRef]

132. Ciambella, J.; Bezazi, A.; Saccomandi, G.; Scarpa, F. Nonlinear elasticity of auxetic open cell foams modeled as continuum solids. *J. Appl. Phys.* **2015**, *117*, 184902. [CrossRef]

133. Lee, J.; Choi, J.B.; Choi, K. Application of homogenization FEM analysis to regular and re-entrant honeycomb structures. *J. Mater. Sci.* **1996**, *31*, 4105–4110. [CrossRef]

134. Huang, F.-Y.; Yan, B.-H.; Yang, D.U. The effects of material constants on the micropolar elastic honeycomb structure with negative Poisson's ratio using the finite element method. *Eng. Comput.* **2002**, *19*, 742–763. [CrossRef]

135. Yang, D.U.; Lee, S.; Huang, F.Y. Geometric effects on micropolar elastic honeycomb structure with negative Poisson's ratio using the finite element method. Finite Elem. *Anal. Des.* **2003**, *39*, 187–205.

136. Liu, W.; Wang, N.; Huang, J.; Zhong, H. The effect of irregularity, residual convex units and stresses on the effective mechanical properties of 2D auxetic cellular structure. *Mater. Sci. Eng. A* **2014**, *609*, 26–33. [CrossRef]

137. Yang, C.; Vora, H.D.; Chang, Y.B. Application of Auxetic Polymeric Structures for Body Protection. In Proceedings of the ASME 2016 Conference on Smart Materials, Adaptive Structures & *Intelligent Systems*, Stowe, VT, USA, 28–30 September 2016; pp. 1–5.

138. Strek, T.; Jopek, H.; Idczak, E.; Wojciechowski, K.W. Computational modelling of structures with non-intuitive behaviour. *Materials* **2017**, *10*, 1386. [CrossRef] [PubMed]

139. Bezazi, A.; Scarpa, F.; Remillat, C. A novel centresymmetric honeycomb composite structure. *Compos. Struct.* **2005**, *71*, 356–364. [CrossRef]

140. Mousanezhad, D.; Ebrahimi, H.; Haghpanah, B.; Ghosh, R.; Ajdari, A.; Hamouda, A.M.S.; Vaziri, A. Spiderweb honeycombs. *Int. J. Solids Struct.* **2015**, *66*, 218–227. [CrossRef]

141. Grima, J.N.; Cauchi, R.; Gatt, R.; Attard, D. Honeycomb composites with auxetic out-of-plane characteristics. *Compos. Struct.* **2013**, *106*, 150–159. [CrossRef]

142. Silva, T.A.A.; Panzera, T.H.; Brandão, L.C.; Lauro, C.H.; Boba, K.; Scarpa, F. Preliminary investigations on auxetic structures based on recycled rubber. *Phys. Status Solidi Basic Res.* **2012**, *249*, 1353–1358. [CrossRef]

143. Milton, G.W. Composite materials with poisson's ratios close to—1. *J. Mech. Phys. Solids* **1992**, *40*, 1105–1137. [CrossRef]

144. Evans, K.E.; Nkansah, M.A.; Hutchinson, I.J. Modelling Negative Poisson Ratio Effects in Network-Embedded Composites. *Acta Metall. Mater.* **1992**, *40*, 2463–2469. [CrossRef]

145. Strek, T.; Jopek, H.; Maruszewski, B.T.; Nienartowicz, M. Computational analysis of sandwich-structured composites with an auxetic phase. *Phys. Status Solidi Basic Res.* **2014**, *251*, 354–366. [CrossRef]

146. Strek, T.; Jopek, H.; Nienartowicz, M. Dynamic response of sandwich panels with auxetic cores. *Phys. Status Solidi Basic Res.* **2015**, *252*, 1540–1550. [CrossRef]

147. Poźniak, A.A.; Wojciechowski, K.W.; Grima, J.N.; Mizzi, L. Planar auxeticity from elliptic inclusions. *Compos. Part B Eng.* **2016**, *94*, 379–388. [CrossRef]

148. Strek, T.; Jopek, H.; Idczak, E. Computational design of two-phase auxetic structures. *Phys. Status Solidi Basic Res.* **2016**, *253*, 1387–1394. [CrossRef]

149. Jopek, H.; Stręk, T. Thermoauxetic behavior of composite structures. *Materials* **2018**, *11*, 294. [CrossRef] [PubMed]

150. Jopek, H.; Strek, T. Thermal and structural dependence of auxetic properties of composite materials. *Phys. Status Solidi Basic Res.* **2015**, *252*, 1551–1558. [CrossRef]

151. Doyoyo, M.; Wan Hu, J. Plastic failure analysis of an auxetic foam or inverted strut lattice under longitudinal and shear loads. *J. Mech. Phys. Solids* **2006**, *54*, 1479–1492. [CrossRef]

152. Qiao, J.X.; Chen, C.Q. Impact resistance of uniform and functionally graded auxetic double arrowhead honeycombs. *Int. J. Impact Eng.* **2015**, *83*, 47–58. [CrossRef]

153. Qiao, J.; Chen, C.Q. Analyses on the In-Plane Impact Resistance of Auxetic Double Arrowhead Honeycombs. *J. Appl. Mech.* **2015**, *82*, 51007. [CrossRef]

154. Lu, Z.; Wang, Q.; Li, X.; Yang, Z. Elastic properties of two novel auxetic 3D cellular structures. *Int. J. Solids Struct.* **2017**, *124*, 46–56. [CrossRef]

155. Shufrin, I.; Pasternak, E.; Dyskin, A.V. Negative Poisson's ratio in hollow sphere materials. *Int. J. Solids Struct.* **2015**, *54*, 192–214. [CrossRef]

156. Spadoni, A.; Ruzzene, M. Elasto-static micropolar behavior of a chiral auxetic lattice. *J. Mech. Phys. Solids* **2012**, *60*, 156–171. [CrossRef]

157. Qi, C.; Remennikov, A.; Pei, L.Z.; Yang, S.; Yu, Z.H.; Ngo, T.D. Impact and close-in blast response of auxetic honeycomb-cored sandwich panels: Experimental tests and numerical simulations. *Compos. Struct.* **2017**, *180*, 161–178. [CrossRef]

158. Wang, X.T.; Wang, B.; Li, X.W.; Ma, L. Mechanical properties of 3D re-entrant auxetic cellular structures. *Int. J. Mech. Sci.* **2017**, *131*, 396–407. [CrossRef]

159. Ranga, D.; Strangwood, M. Finite element modelling of the quasi-static and dynamic behaviour of a solid sports ball based on component material properties. *Procedia Eng.* **2010**, *2*, 3287–3292. [CrossRef]

160. Jiang, L.; Hu, H. Low-velocity impact response of multilayer orthogonal structural composite with auxetic effect. *Compos. Struct.* **2017**, *169*, 62–68. [CrossRef]

161. Miller, W.; Smith, C.W.; Scarpa, F.; Evans, K.E. Flatwise buckling optimization of hexachiral and tetrachiral honeycombs. *Compos. Sci. Technol.* **2010**, *70*, 1049–1056. [CrossRef]

162. Novak, N.; Vesenjak, M.; Ren, Z. Computational Simulation and Optimization of Functionally Graded Auxetic Structures Made from Inverted Tetrapods. *Phys. Status Solidi Basic Res.* **2017**, *254*, 1–7. [CrossRef]

163. Safikhani Nasim, M.; Etemadi, E. Three dimensional modeling of warp and woof periodic auxetic cellular structure. *Int. J. Mech. Sci.* **2018**, *136*, 475–481. [CrossRef]

164. Carta, G.; Brun, M.; Baldi, A. Design of a porous material with isotropic negative Poisson's ratio. *Mech. Mater.* **2016**, *97*, 67–75. [CrossRef]

165. Scarpa, F.; Blain, S.; Lew, T.; Perrott, D.; Ruzzene, M.; Yates, J.R. Elastic buckling of hexagonal chiral cell honeycombs. *Compos. Part A Appl. Sci. Manuf.* **2007**, *38*, 280–289. [CrossRef]

166. Wang, Z.; Hu, H. A finite element analysis of an auxetic warp-knitted spacer fabric structure. *Text. Res. J.* **2015**, *85*, 404–415. [CrossRef]

167. Gao, Q.; Wang, L.; Zhou, Z.; Ma, Z.D.; Wang, C.; Wang, Y. Theoretical, numerical and experimental analysis of three-dimensional double-V honeycomb. *Mater. Des.* **2018**, *139*, 380–391. [CrossRef]

168. Farhan, M.; Shahid, M. Negative Poissons ratio behavior of idealized elastomeric auxetic cellular structures for various carbon black nanoparticles loadings. *J. Elastomers Plast.* **2015**, *47*, 479–487. [CrossRef]

169. Javadi, A.A.; Faramarzi, A.; Farmani, R. Design and optimization of microstructure of auxetic materials. *Eng. Comput.* **2012**, *29*, 260–276. [CrossRef]

170. Lee, D.; Shin, S. Evaluation of Optimized Topology Design of Cross-Formed Structures with a Negative Poisson's Ratio. *Iran. J. Sci. Technol. Trans. Civ. Eng.* **2016**, *40*, 109–120. [CrossRef]

171. Vogiatzis, P.; Chen, S.; Wang, X.; Li, T.; Wang, L. Topology optimization of multi-material negative Poisson's ratio metamaterials using a reconciled level set method. *Comput. Aided Des.* **2017**, *83*, 15–32. [CrossRef]

172. Ren, X.; Shen, J.; Tran, P.; Ngo, T.D.; Xie, Y.M. Design and characterisation of a tuneable 3D buckling-induced auxetic metamaterial. *Mater. Des.* **2018**, *139*, 336–342. [CrossRef]

173. Li, H.; Luo, Z.; Gao, L.; Walker, P. Topology optimization for functionally graded cellular composites with metamaterials by level sets. *Comput. Methods Appl. Mech. Eng.* **2017**, *328*, 340–364. [CrossRef]

174. Qin, H.; Yang, D.; Ren, C. Modelling theory of functional element design for metamaterials with arbitrary negative Poisson's ratio. *Comput. Mater. Sci.* **2018**, *150*, 121–133. [CrossRef]

175. Yang, S.; Qi, C.; Guo, D.M.; Wang, D. Energy absorption of an re-entrant honeycombs with negative Poisson's ratio. *Appl. Mech. Mater.* **2012**, *148*, 992–995. [CrossRef]

176. Hou, X.; Deng, Z.; Zhang, K. Dynamic Crushing Strength Analysis of Auxetic Honeycombs. *Acta Mech. Solida Sin.* **2016**, *29*, 490–501. [CrossRef]

177. Wang, Y.; Wang, L.; Ma, Z.D.; Wang, T. A negative Poisson's ratio suspension jounce bumper. *Mater. Des.* **2016**, *103*, 90–99. [CrossRef]

178. Imbalzano, G.; Tran, P.; Ngo, T.D.; Lee, P.V.S. Three-dimensional modelling of auxetic sandwich panels for localised impact resistance. *J. Sandw. Struct. Mater.* **2017**, *19*, 291–316. [CrossRef]

179. Imbalzano, G.; Tran, P.; Ngo, T.D.; Lee, P.V.S. A numerical study of auxetic composite panels under blast loadings. *Compos. Struct.* **2016**, *135*, 339–352. [CrossRef]

180. Critchley, R.; Corni, I.; Stokes, K.; Walsh, F.C.; Wharton, J.; Wood, R.J.K. High Strain Materials for Body Armour Inspired from Nature High Strain Materials for Body Armour Inspired from Nature. 2014. Available online: https://www.researchgate.net/publication/267261907 (accessed on 4 June 2018).

181. Bryson, A.; Smith, L. Impact response of sports materials. *Procedia Eng.* **2010**, *2*, 2961–2966. [CrossRef]

182. Mills, N.J.; Gilchrist, A. Oblique impact testing of bicycle helmets. *Int. J. Impact Eng.* **2008**, *35*, 1075–1086. [CrossRef]

183. Scarpa, F.; Giacomin, J.; Zhang, Y.; Pastorino, P. Mechanical performance of auxetic polyurethane foam for antivibration glove applications. *Cell. Polym.* **2005**, *24*, 253–268.

184. Choi, J.B.; Lakes, R.S. Nonlinear Analysis of the Poisson's Ratio of Negative Poisson's Ratio Foams. *J. Compos. Mater.* **1994**, *29*, 113–128. [CrossRef]

185. Chan, N.; Evans, K.E. Fabrication methods for auxetic foams. *J. Mater. Sci.* **1997**, *32*, 5945–5953. [CrossRef]

186. Loureiro, M.A.; Lakes, R.S. Scale-up of transformation of negative Poisson's ratio foam: Slabs. *Cell. Polym.* **1997**, *16*, 349–363.

187. Bianchi, M.; Scarpa, F.; Banse, M.; Smith, C.W. Novel generation of auxetic open cell foams for curved and arbitrary shapes. *Acta Mater.* **2011**, *59*, 686–691. [CrossRef]

188. Allen, T.; Hewage, T.; Newton-Mann, C.; Wang, W.; Duncan, O.; Alderson, A. Fabrication of Auxetic Foam Sheets for Sports Applications. *Phys. Status Solidi Basic Res.* **2017**, *254*, 1700596. [CrossRef]

189. Choi, J.B.; Lakes, R.S. Nonlinear properties of polymer cellular materials with a negative Poisson's ratio. *Mater. Sci.* **1992**, *27*, 4678–4684. [CrossRef]

190. Alderson, A.; Davies, P.J.; Alderson, K.; Smart, G.M. The Effects of Processing on the Topology and Mechanical Properties of Negative Poisson's Ratio Foams. In Proceedings of the 2005 ASME International Mechanical Engineering Congress and Exposition, Orlando, FL, USA, 5–11 November 2005; pp. 1–8.

191. Gatt, R.; Attard, D.; Manicaro, E.; Chetcuti, E.; Grima, J.N. On the effect of heat and solvent exposure on the microstructure properties of auxetic foams: A preliminary study. *Phys. Status Solidi Basic Res.* **2011**, *248*, 39–44. [CrossRef]

192. Bianchi, M.; Frontoni, S.; Scarpa, F.; Smith, C.W. Density change during the manufacturing process of PU-PE open cell auxetic foams. *Phys. Status Solidi Basic Res.* **2011**, *248*, 30–38. [CrossRef]

193. Scarpa, F.; Pastorino, P.; Garelli, A.; Patsias, S.; Ruzzene, M. Auxetic compliant flexible PU foams: Static and dynamic properties. *Phys. Status Solidi Basic Res.* **2005**, *242*, 681–694. [CrossRef]

194. Duncan, O.; Allen, T.; Foster, L.; Gatt, R.; Grima, J.N.; Alderson, A. Controlling Density and Modulus in Auxetic Foam Fabrications—Implications for Impact and Indentation Testing. *Proceedings* **2018**, *2*, 250. [CrossRef]

195. Boba, K.; Bianchi, M.; McCombe, G.; Gatt, R.; Griffin, A.C.; Richardson, R.M.; Scarpa, F.; Hamerton, I.; Grima, J. Blocked shape memory effect in negative Poisson's ratio polymer metamaterials. *ACS Appl. Mater. Interfaces* **2016**, *8*, 20319–20328. [CrossRef] [PubMed]

196. Grima, J.N.; Attard, D.; Gatt, R.; Cassar, R.N. A novel process for the manufacture of auxetic foams and for their re-conversion to conventional form. *Adv. Eng. Mater.* **2009**, *11*, 533–535. [CrossRef]

197. Lowe, A.; Lakes, R.S. Negative Poisson's ratio foam as seat cushion material. *Cell. Polym.* **2000**, *19*, 157–167.

198. Chan, N.; Evans, K.E. Microscopic examination of the microstructure and deformation of conventional and auxetic foams. *J. Mater. Sci.* **1997**, *2*, 5725–5736. [CrossRef]

199. Martz, E.O.; Lee, T.; Lakes, R.S.; Goel, V.K.; Park, J.B. Re-entrant transformation methods in closed cell foams. *Cell. Polym.* **1996**, *15*, 229–249.

200. Quadrini, F.; Bellisario, D.; Ciampoli, L.; Costanza, G.; Santo, L. Auxetic epoxy foams produced by solid state foaming. *J. Cell. Plast.* **2015**, *52*, 441–454. [CrossRef]

201. Xu, T.; Li, G. A shape memory polymer based syntactic foam with negative Poisson' s ratio. *Mater. Sci. Eng. A* **2011**, *528*, 6804–6811. [CrossRef]

202. Wang, K.; Chang, Y.H.; Chen, Y.; Zhang, C.; Wang, B. Designable dual-material auxetic metamaterials using three-dimensional printing. *Mater. Des.* **2015**, *67*, 159–164. [CrossRef]

203. McDonald, S.A.; Ravirala, N.; Withers, P.J.; Alderson, A. In situ three-dimensional X-ray microtomography of an auxetic foam under tension. *Scr. Mater.* **2009**, *60*, 232–235. [CrossRef]

204. Lisiecki, J.; Klysz, S.; Blazejewicz, T.; Gmurczyk, G.; Reymer, P. Tomographic examination of auxetic polyurethane foam structures. *Phys. Status Solidi Basic Res.* **2013**, *251*, 314–320. [CrossRef]

205. American Society for Testing and Materials. *ASTM-D412-15a: Standard Test Methods for Vulcanized Rubber and Thermoplastic Elastomers—Tension*; ASTM International: West Conshohocken, PA, USA, 2015; Volume 9, pp. 1–14.

206. Miller, W.; Hook, P.B.; Smith, C.W.; Wang, X.; Evans, K.E. The manufacture and characterisation of a novel, low modulus, negative Poisson's ratio composite. *Compos. Sci. Technol.* **2009**, *69*, 651–655. [CrossRef]

207. Miller, W.; Ren, Z.; Smith, C.W.; Evans, K.E. A negative Poisson's ratio carbon fibre composite using a negative Poisson's ratio yarn reinforcement. *Compos. Sci. Technol.* **2012**, *72*, 761–766. [CrossRef]

208. Bhattacharya, S.; Zhang, G.H.; Ghita, O.; Evans, K.E. The variation in Poisson's ratio caused by interactions between core and wrap in helical composite auxetic yarns. *Compos. Sci. Technol.* **2014**, *102*, 87–93. [CrossRef]

209. Zhang, G.; Ghita, O.R.; Evans, K.E. Dynamic thermo-mechanical and impact properties of helical auxetic yarns. *Compos. Part B Eng.* **2016**, *99*, 494–505. [CrossRef]

210. Zhang, G.H.; Ghita, O.; Evans, K.E. The fabrication and mechanical properties of a novel 3-component auxetic structure for composites. *Compos. Sci. Technol.* **2015**, *117*, 257–267. [CrossRef]

211. Zhang, G.; Ghita, O.R.; Lin, C.; Evans, K.E. Large-scale manufacturing of helical auxetic yarns using a novel semi-coextrusion process. *Text. Res. J.* **2017**, 1–12. [CrossRef]

212. Ge, Z.; Hu, H.; Liu, S. A novel plied yarn structure with negative Poisson's ratio. *J. Text. Inst.* **2016**, *107*, 578–588. [CrossRef]

213. Lim, T.C. Semi-auxetic yarns. *Phys. Status Solidi Basic Res.* **2014**, *251*, 273–280. [CrossRef]

214. Liu, S.; Pan, X.; Zheng, D.; Du, Z.; Liu, G.; Yang, S. Study on the structure formation and heat treatment of helical auxetic complex yarn. *Text. Res. J.* **2018**. [CrossRef]

215. Sloan, M.R.; Wright, J.R.; Evans, K.E. The helical auxetic yarn—A novel structure for composites and textiles; Geometry, manufacture and mechanical properties. *Mech. Mater.* **2011**, *43*, 476–486. [CrossRef]

216. McAfee, J.; Faisal, N.H. Parametric sensitivity analysis to maximise auxetic effect of polymeric fibre based helical yarn. *Compos. Struct.* **2017**, *162*, 1–12. [CrossRef]

217. Wright, J.R.; Burns, M.K.; James, E.; Sloan, M.R.; Evans, K.E. On the design and characterisation of low-stiffness auxetic yarns and fabrics. *Text. Res. J.* **2012**, *82*, 645–654. [CrossRef]

218. Alderson, K.L.; Alderson, A.; Smart, G.; Simkins, V.R.; Davies, P.J. Auxetic polypropylene fibres: Part 1—Manufacture and characterisation. *Plast Rubber Compos.* **2002**, *31*, 344–349. [CrossRef]

219. Ravirala, N.; Alderson, A.; Alderson, K.L.; Davies, P.J. Auxetic polypropylene films. *Polym. Eng. Sci.* **2005**, *45*, 517–528. [CrossRef]

220. Alderson, A.; Alderson, K. Expanding materials and applications: Exploiting auxetic textiles. *Tech. Text. Int.* **2005**, *14*, 29–34.

221. Simkins, V.R.; Alderson, A.; Davies, P.J.; Alderson, K.L. Single fibre pullout tests on auxetic polymeric fibres. *J. Mater. Sci.* **2005**, *40*, 4355–4364. [CrossRef]

222. Ng, W.S.; Hu, H. Woven fabrics made of auxetic plied yarns. *Polymers* **2018**, *10*, 226.

223. Alderson, A. *Smart Solutions from Auxetic Materials*; Med-Tech Innovation: Chester, UK, 2011.

224. Verma, P.; Shofner, M.L.; Lin, A.; Wagner, K.B.; Griffin, A.C. Induction of auxetic response in needle-punched nonwovens: Effects of temperature, pressure, and time. *Phys. Status Solidi* **2016**, *253*, 1270–1278. [CrossRef]

225. Verma, P.; Shofner, M.L.; Lin, A.; Wagner, K.B.; Griffin, A.C. Inducing out-of-plane auxetic behavior in needle-punched nonwovens. *Phys. Status Solidi Basic Res.* **2015**, *252*, 1455–1464. [CrossRef]

226. Zulifqar, A.; Hua, T.; Hu, H. Development of UNI-stretch woven fabrics with zero and negative Poisson's ratio. *Text. Res. J.* **2017**. [CrossRef]

227. Alderson, K.; Alderson, A.; Anand, S.; Simkins, V.; Nazare, S.; Ravirala, N. Auxetic warp knit textile structures. *Phys. Status Solidi Basic Res.* **2012**, *249*, 1322–1329. [CrossRef]

228. Ma, P.; Chang, Y.; Jiang, G. Design and fabrication of auxetic warp-knitted structures with a rotational hexagonal loop. *Text. Res. J.* **2016**, *86*, 2151–2157. [CrossRef]

229. Ugbolue, S.C.; Kim, Y.K.; Warner, S.B.; Fan, Q.; Yang, C.L.; Kyzymchuk, O.; Feng, Y.; Lord, J. The formation and performance of auxetic textiles. Part II: Geometry and structural properties. *J. Text. Inst.* **2011**, *102*, 424–433. [CrossRef]

230. Wang, Z.; Hu, H.; Xiao, X. Deformation behaviors of three-dimensional auxetic spacer fabrics. *Text. Res. J.* **2014**, *84*, 1361–1372. [CrossRef]

231. Wang, Z.; Hu, H. Tensile and forming properties of auxetic warp-knitted spacer fabrics. *Text. Res. J.* **2017**, *87*, 1925–1937. [CrossRef]

232. Chang, Y.; Ma, P. Fabrication and property of auxetic warp-knitted spacer structures with mesh. *Text. Res. J.* **2017**. [CrossRef]

233. Chang, Y.; Ma, P.; Jiang, G. Energy absorption property of warp-knitted spacer fabrics with negative Possion's ratio under low velocity impact. *Compos. Struct.* **2017**, *182*, 471–477. [CrossRef]

234. Chang, Y.; Ma, P. Energy absorption and Poisson's ratio of warp-knitted spacer fabrics under uniaxial tension. *Text. Res. J.* **2018**. [CrossRef]

235. Hu, H.; Wang, Z.; Liu, S. Development of auxetic fabrics using flat knitting technology. *Text. Res. J.* **2011**, *81*, 1493–1502.

236. Steffens, F.; Oliveira, F.R.; Mota, C.; Fangueiro, R. High-performance composite with negative Poisson's ratio. *J. Mater. Res.* **2017**, *32*, 3477–3484. [CrossRef]

237. Goto, K.; Arai, M.; Matsuda, T.; Kubo, G. Elasto-viscoplastic analysis for negative through-the-thickness Poisson's ratio of woven laminate composites based on homogenization theory. *Int. J. Mech. Sci.* **2017**. [CrossRef]

238. Ge, Z.; Hu, H.; Liu, Y. Numerical analysis of deformation behavior of a 3D textile structure with negative Poisson's ratio under compression. *Text. Res. J.* **2015**, *85*, 548–557. [CrossRef]

239. Ge, Z.; Hu, H. A theoretical analysis of deformation behavior of an innovative 3D auxetic textile structure. *J. Text. Inst.* **2015**, *106*, 101–109. [CrossRef]

240. Nava-Gómez, G.G.; Camacho-Montes, H.; Sabina, F.J.; Rodríguez-Ramos, R.; Fuentes, L.; Guinovart-Díaz, R. Elastic properties of an orthotropic binary fiber-reinforced composite with auxetic and conventional constituents. *Mech. Mater.* **2012**, *48*, 1–25. [CrossRef]

241. Liaqat, M.; Samad, H.A.; Hamdani, S.T.A.; Nawab, Y. The development of novel auxetic woven structure for impact applications. *J. Text. Inst.* **2017**, *108*, 1264–1270. [CrossRef]

242. American Society for Testing and Materials. *D6110-17: Standard Test Method for Determining the Charpy Impact Resistance of Notched Specimens of Plastics*; ASTM International: West Conshohocken, PA, USA, 2010; pp. 1–17.

243. Zhou, L.; Jiang, L.; Hu, H. Auxetic composites made of 3D textile structure and polyurethane foam. *Phys. Status Solidi Basic Res.* **2016**, *257*, 1331–1341. [CrossRef]

244. Jiang, L.; Gu, B.; Hu, H. Auxetic composite made with multilayer orthogonal structural reinforcement. *Compos. Struct.* **2016**, *135*, 23–29. [CrossRef]

245. Wang, Z.; Hu, H. Auxetic materials and their potential applications in textiles. *Text. Res. J.* **2014**, *84*, 1600–1611. [CrossRef]

246. Ma, P.; Chang, Y.; Boakye, A.; Jiang, G. Review on the knitted structures with auxetic effect. *J. Text. Inst.* **2017**, *108*, 947–961. [CrossRef]

247. Yuan, S.; Shen, F.; Bai, J.; Chua, C.K.; Wei, J.; Zhou, K. 3D soft auxetic lattice structures fabricated by selective laser sintering: TPU powder evaluation and process optimization. *Mater. Des.* **2017**, *120*, 317–327. [CrossRef]

248. Lantada, A.D.; De Blas Romero, A.; Schwentenwein, M.; Jellinek, C.; Homa, J. Lithography-based ceramic manufacture (LCM) of auxetic structures: Present capabilities and challenges. *Smart Mater. Struct.* **2016**, *25*, 054015. [CrossRef]

249. Critchley, B.R.; Corni, I.; Wharton, J.A.; Walsh, F.C.; Wood, R.J.K.; Stokes, K.R. The Preparation of Auxetic Foams by Three-Dimensional Printing and Their Characteristics. *Adv. Eng. Mater.* **2013**, *15*, 980–985. [CrossRef]

250. Hengsbach, S.; Lantada, A.D. Direct laser writing of auxetic structures: Present capabilities and challenges. *Smart Mater. Struct.* **2014**, *23*, 085033. [CrossRef]

251. Yang, L.; Harrysson, O.; West, H.; Cormier, D. Mechanical properties of 3D re-entrant honeycomb auxetic structures realized via additive manufacturing. *Int. J. Solids Struct.* **2015**, *69*, 475–490. [CrossRef]

252. Li, T.; Wang, L. Bending behavior of sandwich composite structures with tunable 3D-printed core materials. *Compos. Struct.* **2017**, *175*, 46–57. [CrossRef]

253. Li, D.; Ma, J.; Dong, L.; Lakes, R.S. Stiff square structure with a negative Poisson's ratio. *Mater. Lett.* **2017**, *188*, 149–151. [CrossRef]

254. Ma, Q.; Peel, L.D. Development of spiral auxetic structures. *Compos. Struct.* **2018**, *192*, 310–316. [CrossRef]

255. Boldrin, L.; Hummel, S.; Scarpa, F.; Di Maio, D.; Lira, C.; Ruzzene, M.; Remillat, C.; Lim, T.; Rajasekaran, R.; Patsias, S. Dynamic behaviour of auxetic gradient composite hexagonal honeycombs. *Compos. Struct.* **2016**, *149*, 114–124. [CrossRef]

256. Huang, H.H.; Wong, B.L.; Chou, Y.C. Design and properties of 3D-printed chiral auxetic metamaterials by reconfigurable connections. *Phys. Status Solidi Basic Res.* **2016**, *253*, 1557–1564. [CrossRef]

257. Saxena, K.K.; Calius, E.P.; Das, R. Tailoring Cellular Auxetics For Wearable Applications with Multimaterial 3D Printing. In Proceedings of the ASME 2016 International Mechanical Engineering Congress & Exposition, Phoenix, AZ, USA, 11–17 November 2016.

258. Saxena, K.K.; Das, R.; Calius, E.P. 3D printable multimaterial cellular auxetics with tunable stiffness. *arXiv* **2017**, arXiv:1707.04486.

259. Jiang, Y.; Li, Y. Novel 3D-Printed Hybrid Auxetic Mechanical Metamaterial with Chirality-Induced Sequential Cell Opening Mechanisms. *Adv. Eng. Mater.* **2018**, *20*, 1–9. [CrossRef]

260. Jiang, Y.; Li, Y. 3D Printed Auxetic Mechanical Metamaterial with Chiral Cells and Re-entrant Cores. *Sci. Rep.* **2018**, *8*, 2397. [CrossRef] [PubMed]

261. Brighenti, R. Smart behaviour of layered plates through the use of auxetic materials. *Thin-Walled Struct.* **2014**, *84*, 432–442. [CrossRef]

262. Ingrole, A.; Hao, A.; Liang, R. Design and modeling of auxetic and hybrid honeycomb structures for in-plane property enhancement. *Mater. Des.* **2017**, *117*, 72–83. [CrossRef]

263. Li, X.; Lu, Z.; Yang, Z.; Yang, C. Directions dependence of the elastic properties of a 3D augmented re-entrant cellular structure. *Mater. Des.* **2017**, *134*, 151–162. [CrossRef]

264. Zhang, X.; Yang, D. Mechanical properties of auxetic cellular material consisting of re-entrant hexagonal honeycombs. *Materials* **2016**, *9*, 900. [CrossRef] [PubMed]

265. Li, D.; Dong, L.; Lakes, R.S. A unit cell structure with tunable Poisson's ratio from positive to negative. *Mater. Lett.* **2016**, *164*, 456–459. [CrossRef]

266. Ebrahimi, H.; Mousanezhad, D.; Nayeb-Hashemi, H.; Norato, J.; Vaziri, A. 3D cellular metamaterials with planar anti-chiral topology. *Mater. Des.* **2018**, *145*, 226–231. [CrossRef]

267. Fu, M.; Liu, F.; Hu, L. A novel category of 3D chiral material with negative Poisson's ratio. *Compos. Sci. Technol.* **2018**, *160*, 111–118. [CrossRef]

268. Hanifpour, M.; Petersen, C.F.; Alava, M.J.; Zapperi, S. Mechanics of disordered auxetic metamaterials. *arXiv* **2017**, arXiv:1704.00943.

269. Yang, C.; Vora, H.D.; Chang, Y.B. Evaluation of Auxetic Polymeric Structures for Use in Protective Pads. In Proceedings of the ASME 2016 International Mechanical Engineering Congress & Exposition, Phoenix, AZ, USA, 11–17 November 2016.

270. Yang, C.; Vora, H.; Chang, Y. Behavior of Auxetic Structures Under Compression and Impact Forces. *Smart Mater. Struct.* **2017**, *27*, 025012. [CrossRef]

271. Bates, S.R.G.; Farrow, I.R.; Trask, R.S. 3D printed polyurethane honeycombs for repeated tailored energy absorption. *Mater. Des.* **2016**, *112*, 172–183. [CrossRef]

272. Alderson, A.; Alderson, K.L.; McDonald, S.A.; Mottershead, B.; Nazare, S.; Withers, P.J.; Yao, Y. Piezomorphic materials. *Macromol. Mater. Eng.* **2013**, *298*, 318–327. [CrossRef]

273. Zorzetto, L.; Ruffoni, D. Re-entrant inclusions in cellular solids: From defects to reinforcements. *Compos. Struct.* **2017**, *176*, 195–204. [CrossRef]

274. Lakes, R.S.; Elms, K. Indentability of conventional and negative Poisson's ratio foams. *J. Compos. Mater.* **1993**, *27*, 1193–1202. [CrossRef]

275. Scarpa, F.; Yates, J.R.; Ciffo, L.G.; Patsias, S. Dynamic crushing of auxetic open-cell polyurethane foam. *Proc. Inst. Mech. Eng. Part C* **2002**, *216*, 1153–1156. [CrossRef]

276. Bocquet, C. Bra with Variable-Volume Straps. WO Patent 2006045935A, 4 May 2006.

277. Timishenko, S.P.; Goodier, J.N. *Theory of Elasticity*, 3rd ed.; McGraw-Hill: New York, NY, USA, 1970.

278. Budynas. *R.G. Advanced Strength and Applied Stress Analysis*; WCB/McGraw-Hill: New York, NY, USA, 1999; pp. 21, 80.

279. Roark, R.J.; Young, W.C. *Formulas for Stress and Strain*; McGraw-Hill: New York, NY, USA, 2012; pp. 20–22, 48–50.

280. Lempriere, B.M. Poisson's ratio in orthotropic materials. *AIAA J.* **1968**, *6*, 2226–2227. [CrossRef]

281. Wojciechowski, K.W. Remarks on "Poisson ratio beyond the limits of the elasticity theory". *J. Phys. Soc. Jpn.* **2003**, *72*, 1819–1820. [CrossRef]

282. Lakes, R.S. Design Considerations for Materials with Negative Poisson's Ratios. *J. Mech. Des.* **1993**, *115*, 696. [CrossRef]

283. Alderson, K.L.; Fitzgerald, A.; Evans, K.E. The strain dependent indentation resilience of auxetic microporous polyethylene. *J. Mater. Sci.* **2000**, *35*, 4039–4047. [CrossRef]

284. Lin, D.C.; Shreiber, D.I.; Dimitriadis, E.K.; Horkay, F. Spherical indentation of soft matter beyond the Hertzian regime: Numerical and experimental validation of hyperelastic models. *Biomech. Model. Mechanobiol.* **2009**, *8*, 345–358. [CrossRef] [PubMed]

285. Argatov, I.I.; Guinovart-Díaz, R.; Sabina, F.J. On local indentation and impact compliance of isotropic auxetic materials from the continuum mechanics viewpoint. *Int. J. Eng. Sci.* **2012**, *54*, 42–57. [CrossRef]

286. Wang, Y.C.; Lakes, R. Analytical parametric analysis of the contact problem of human buttocks and negative Poisson's ratio foam cushions. *Int. J. Solids Struct.* **2002**, *39*, 4825–4838. [CrossRef]

287. Waters, N.E. The indentation of thin rubberer sheets by cylindrical indetors. *Br. J. Appl. Phys.* **1965**, *16*, 1387–1392. [CrossRef]

288. Photiou, D.; Sarris, E.; Constantinides, G. On the conical indentation response of elastic auxetic materials: Effects of Poisson's ratio, contact friction and cone angle. *Int. J. Solids Struct.* **2017**, *1*, 33–42. [CrossRef]

289. Photiou, D.; Sarris, E.; Constantinides, G. Erratum to "On the conical indentation response of elastic auxetic materials: Effects of Poisson's ratio, contact friction and cone angle" [Int. J. Solids Struct. 81 (2016) 33–42]. *Int. J. Solids Struct.* **2017**, *110–111*, 404. [CrossRef]

290. Guo, X.; Jin, F.; Gao, H. Mechanics of non-slipping adhesive contact on a power-law graded elastic half-space. *Int. J. Solids Struct.* **2011**, *48*, 2565–2575. [CrossRef]

291. Li, S.; Al-Badani, K.; Gu, Y.; Lake, M.; Li, L.; Rothwell, G.; Ren, J. The Effects of Poisson's Ratio on the Indentation Behavior of Materials with Embedded System in an Elastic Matrix. *Phys. Status Solidi Basic Res.* **2017**, *254*, 1–8. [CrossRef]

292. Aw, J.; Zhao, H.; Norbury, A.; Li, L.; Rothwell, G.; Ren, J. Effects of Poisson's ratio on the deformation of thin membrane structures under indentation. *Phys. Status Solidi Basic Res.* **2015**, *252*, 1526–1532. [CrossRef]

293. Morris, D.J.; Cook, R.F. Indentation fracture of low-dielectric constant films: Part II. Indentation fracture mechanics model. *J. Mater. Res.* **2008**, *23*, 2429–2442. [CrossRef]

294. Argatov, I.I.; Sabina, F.J. Small-scale indentation of an elastic coated half-space: The effect of compliant substrate. *Int. J. Eng. Sci.* **2016**, *104*, 87–96. [CrossRef]

295. Chan, N.; Evans, K.E. The mechanical properties of conventional and auxetic foams. Part I: Compression and tension. *J. Cell. Plast.* **1999**, *35*, 130–165. [CrossRef]

296. Adam, M.M.; Berger, J.R.; Martin, P.A. Singularities in auxetic elastic bimaterials. *Mech. Res. Commun.* **2013**, *47*, 102–105. [CrossRef]

297. Kwon, K.; Phan, A.V. Symmetric-Galerkin boundary element analysis of the dynamic T-stress for the interaction of a crack with an auxetic inclusion. *Mech. Res. Commun.* **2015**, *69*, 91–96. [CrossRef]

298. Song, F.; Zhou, J.; Xu, X.; Xu, Y.; Bai, Y. Effect of a negative poisson ratio in the tension of ceramics. *Phys. Rev. Lett.* **2008**, *100*, 1–4. [CrossRef] [PubMed]

299. Mohsenizadeh, S.; Alipour, R.; Shokri Rad, M.; Farokhi Nejad, A.; Ahmad, Z. Crashworthiness assessment of auxetic foam-filled tube under quasi-static axial loading. *Mater. Des.* **2015**, *88*, 258–268. [CrossRef]

300. Bezazi, A.; Scarpa, F. Mechanical behaviour of conventional and negative Poisson's ratio thermoplastic polyurethane foams under compressive cyclic loading. *Int. J. Fatigue.* **2007**, *29*, 922–930. [CrossRef]

301. Pastorino, P.; Scarpa, F.; Patsias, S.; Yates, J.R.; Haake, S.J.; Ruzzene, M. Strain rate dependence of stiffness and Poisson's ratio of auxetic open cell PU foams. *Phys. Status Solidi Basic Res.* **2007**, *244*, 955–965. [CrossRef]

302. Nakonieczna, P.; Wierzbicki, Ł.; Śladowska, B.; Leonowicz, M.; Lisiecki, J. Composites with Impact Absorption Ability Based on Shear Thickening Fluids and Auxetic Foams. *Compos. Theory Pract.* **2017**, *2*, 67–72.

303. Qi, C.; Yang, S.; Wang, D.; Yang, L.J. Ballistic resistance of honeycomb sandwich panels under in-plane high-velocity impact. *Sci. World J.* **2013**, *2013*, 892781. [CrossRef] [PubMed]

304. Liu, W.; Wang, N.; Luo, T.; Lin, Z. In-plane dynamic crushing of re-entrant auxetic cellular structure. *Mater. Des.* **2016**, *100*, 84–91. [CrossRef]

305. Yang, S.; Chalivendra, V.B.; Kim, Y.K. Impact behaviour of auxetic Kevlar®/epoxy composites. *IOP Conf. Ser. Mater. Sci. Eng.* **2017**, *254*, 042031. [CrossRef]

306. Yang, S.; Chalivendra, V.B.; Kim, Y.K. Fracture and impact characterization of novel auxetic Kevlar®/Epoxy laminated composites. *Compos. Struct.* **2017**, *168*, 120–129. [CrossRef]

307. Alderson, K.L.; Coenen, V.L. The low velocity impact response of auxetic carbon fibre laminates. *Phys. Status Solidi* **2008**, *245*, 489–496. [CrossRef]

308. Coenen, V.L.; Alderson, K.L. Mechanisms of failure in the static indentation resistance of auxetic carbon fibre laminates. *Phys. Status Solidi Basic Res.* **2011**, *248*, 66–72. [CrossRef]

309. International Organization for Standardization. *Mechanical Vibration and Shock—Hand-Arm Vibration—Method for Measuring the Vibration Transmissibility of Resilient Materials When Loaded by the Hand-Arm System;* ISO 13753:1999; ISO: Geneva, Switzerland, 1999.

310. Chen, C.P.; Lakes, R.S. Dynamic wave dispersion and loss properties of conventional and negative Poisson's ratio polymeric cellular materials. *Cell. Polym.* **1989**, *8*, 343–369.

311. Lisiecki, J.; Nowakowski, D.; Reymer, P. Fatigue Properties of Polyurethane Foams, With Special Emphasis on Auxetic Foams, Used for Helicopter Pilot Seat Cushion Inserts. *Fatigue Aircr. Struct.* **2014**, *1*, 72–78. [CrossRef]

312. Evans, K.E. The design of doubly curved sandwich panels with honeycomb cores. *Compos. Struct.* **1991**, *17*, 95–111. [CrossRef]

313. Alderson, A.; Alderson, K.L.; Chirima, G.; Ravirala, N.; Zied, K.M. The in-plane linear elastic constants and out-of-plane bending of 3-coordinated ligament and cylinder-ligament honeycombs. *Compos. Sci. Technol.* **2010**, *70*, 1034–1041. [CrossRef]

314. Mehta, R. A nose for auxetics. *Mater. World* **2010**, *18*, 9–12.

315. Bianchi, M.; Scarpa, F.; Smith, C.W.; Whittell, G.R. Physical and thermal effects on the shape memory behaviour of auxetic open cell foams. *J. Mater. Sci.* **2010**, *45*, 341–347. [CrossRef]

316. Allen, T.; Martinello, N.; Zampieri, D.; Hewage, T.; Senior, T.; Foster, L.; Alderson, A. Auxetic foams for sport safety applications. *Procedia Eng.* **2015**, *112*, 104–109. [CrossRef]

317. Lim, T.C.; Alderson, A.; Alderson, K.L. Experimental studies on the impact properties of auxetic materials. *Phys. Status Solidi.* **2014**, *251*, 307–313. [CrossRef]

318. D3O. Trust Helmet Pad System [Internet]. Web Page. 2018. Available online: https://www.d3o.com/products/trust-helmet-pad-system/ (accessed on 25 January 2018).

319. Xu, K.; Tan, Y.; Xin, J.H.; Liu, Y.; Lu, C.; Deng, Y.; Han, C.; Hu, H.; Wang, P. A novel impact hardening polymer with negative Poisson's ratio for impact protection. *Mater. Today Commun.* **2015**, *5*, 50–59. [CrossRef]
320. Foster, L.; Peketi, P.; Allen, T.; Senior, T.; Duncan, O.; Alderson, A. Application of Auxetic Foam in Sports Helmets. *Appl. Sci.* **2018**, *8*, 354. [CrossRef]

Article

Functional Elastic Knits Made of Bamboo Charcoal and Quick-Dry Yarns: Manufacturing Techniques and Property Evaluations

Jia-Horng Lin [1,2,3,4,5,6], Chih-Hung He [2], Yu-Tien Huang [4] and Ching-Wen Lou [5,6,7,8,*]

[1] School of Textiles, Tianjin Polytechnic University, Tianjin 300387, China; jhlin@fcu.edu.tw
[2] Department of Fiber and Composite Materials, Feng Chia University, Taichung 40768, Taiwan; chihhung.he@gmail.com
[3] School of Chinese Medicine, China Medical University, Taichung 40402, Taiwan
[4] Department of Fashion Design, Asia University, Taichung 41354, Taiwan; jhlin@fcu.edu.tw
[5] Department of Chemistry and Chemical Engineering, Minjiang University, Fuzhou 350108, China
[6] College of Texitle and Clothing, Qingdao University, Shangdong 266071, China
[7] Graduate Institute of Biotechnology and Biomedical Engineering, Central Taiwan University of Science and Technology, Taichung 40601, Taiwan
[8] Innovation Platform of Intelligent and Energy-Saving Textiles, School of Textiles, Tianjin Polytechnic University, Tianjin 300387, China
* Correspondence: cwlou@ctust.edu.tw, Tel.: +886-4-2451-8672; Fax: +886-4-2451-0871

Academic Editor: Thomas Allen
Received: 2 October 2017; Accepted: 5 December 2017; Published: 11 December 2017

Abstract: Conventional sportswear fabrics are functional textiles that can mitigate the impaired muscles caused by exercises for the wearers, but they can also cause discomfort and skin allergy. This study proposes combining two yarns to form functional composite yarns, by using a twisting or wrapping process. Moreover, a different twist number is used in order to adjust the performance of functional composite yarns. A crochet machine is used to make the functional composite yarns into functional elastic knits that are suitable for use in sportswear. The test results show that, in comparison to the non-processed yarns, using the twisted or wrapped yarns can considerably decrease the water vapor transmission rate of functional elastic knits by 38%, while also improving their far infrared emissivity by 13%, water absorption rate by 39%, and air permeability by 136%. In particular, the functional elastic knits that are made of B-wrapped yarns (bamboo charcoal- wrapped yarns), composed of 20 twists per inch, have the optimal diverse functions.

Keywords: quick-dry yarn; bamboo charcoal yarn; mechanical properties; sportswear textiles; functional composite yarns

1. Introduction

Functional textiles have functionalities that are specifically tailored to the needs of users. The pursuits of comfort, and such health purposes as thermal insulation, moisture wicking, antistatic, and anti-ultraviolet properties, become the mainstream of current studies [1–5]. Functional textiles are commonly used in sports and protection fields; the design of sportswear textiles emphasizes the selection of the materials used. Perspiration is a major factor to consider in regards to comfort. People exercise and sweat, and their garment absorbs the perspiration. If the textiles fail to expel the perspiration efficiently, the wet textiles make the wearers feel cold and unpleasant [6–8]. Therefore, scholars use fibers that have perspiration absorbent and moisture exhaling properties to fabricate sportswear textiles. For example, quick-dry yarns have a cross-shaped transection, which helps to transport sweat from the body to the fabric surface and decreases the accumulation of moisture in clothes. This improves the cooling sensation through the evaporation of sweat from the fabric surface [4]. Bamboo charcoal yarns are composed of multiple bamboo charcoal fibers that emit far infrared rays. They have a specific surface area and micro-porous structure that improves the capillary phenomenon. As a result, quick-dry yarns and bamboo charcoal yarns are pervasively used in sportswear textiles [9–13].

The other important function of sportswear textiles is to protect the wearers from muscle hazards caused by doing the exercises. Common protectors are wrist support, knee support, and back support [14–18]. Conventional protection employs elastic knits to restrain muscles, thereby mitigating an excessively exerted force and fatigue release. However, conventional protectors make the wearers feel uncomfortable and restrain humidity in local areas which leads to skin allergies. Unlike the conventional protectors, compression–stretch materials, made of a considerable number of elastic yarns, can exert a certain pressure onto the muscles. The intense compression provides compression fabrics with a denser structure, but simultaneously limits the perspiration efficacy of the textiles. As a result, the moisture and perspiration are kept in the compression materials, which can cause skin allergic reactions for the wearers. In recent years, there is an innovative research trend using far-infrared textiles to take the load off the muscles. Far infrared ray (FIR) textiles, a new category of functional textiles, contribute to putative health and wellbeing functionality. At the molecular level, FIRs exert strong rotational and vibrational effects, and are biologically beneficial. The other feature of FIR textiles is the enhancement of blood circulation. The local tissues of the human body are activated, which in turn causes dilatation of the blood vessels and facilitates metabolism. The activated cells accelerate the regeneration of impaired tissues and the healing of lesions [10]. To sum up, these materials absorb energy from sunlight and then radiate this energy back onto the body at specified wavelengths. The positive results indicated that FIR textiles outperformed conventional textiles in terms of the FIR emissive efficacy [10]. Functional yarns are specifically designed to improve certain functions, such as UV-cut yarns, FIR emissive yarns, and perspiration absorbent and moisture exhaling yarns. Usually, functional yarns are made of at least one specified function. Functional polymer particles are melted after being heated with a high temperature. The melted particles are ejected through the spinneret under pressure, forming functional fiber bundles. The fiber bundles are then expanded and coated with antistatic agent, and finally formed into functional yarns [13]. The process is highly complex and has a high production cost, which makes mass production impossible. Moreover, adding functional powders results in a high surface roughness of the yarns, causing abrasion against the machine components during the knitting process.

The other method for producing multi-functional yarns is using a wrapping or twisting process to combine different functional yarns [19–23]. The process has a low production cost and is easily manufactured, thereby making mass production feasible. Twisting and wrapping can provide composite yarns with multiple functions, but possibly change the structure and cause the absence of other functions. A previous study examined the relationship with the hygroscopicity of yarns and the twisting process. The test results showed that excessive twisting damaged the yarns' structure and decreased the capillary phenomenon [24]. Another previous study examined how the moisture diffusion coefficient of yarns was related to twisting. The test results indicated that twisting resulted in changes in the yarns' structure. The lower the moisture diffusion coefficient, the lower the capillary phenomenon [25].

Our experimental methodology is as follows. The selected materials facilitate the perspiration that the sports textiles emphasize. In order to compensate for the restriction of twisting or wrapping processes against the functions of yarns, bamboo charcoal yarns and quick-dry yarns are two primary materials that are used to meet the requirement of comfort. Different twist numbers and different manufacturing methods are employed in order to compare the functional composite twisted yarns and functional composite wrapped yarns in terms of functionalities. The twisted yarns and wrapped yarns are separately made into functional elastic knits using a crochet machine. The air permeability, far-infrared emissivity, water absorption rate, and mechanical properties of the knits are evaluated, examining the optimal manufacturing parameters.

2. Materials and Methods

2.1. Materials

Bamboo charcoal yarn (Nan Ya Plastics Corporation, Taipei, Taiwan) has a specification of 75D/72F. Quick-dry yarn (Everest Textile Co., Ltd, Taoyuan, Taiwan) has a specification of 75D/48F. Polyester (PET) filament (Yi Jinn Industrial Co., Ltd., Taipei, Taiwan) has a specification of 150 D/48F. Rubber thread (Ta Yu Co., Ltd., Taichung, Taiwan) has a diameter of 0.65 mm.

2.2. Preparation of Functional Composite Yarns and Functional Elastic Knits

In order to examine the influences of wrapping and twisting processes on the properties of the functional elastic knits, bamboo charcoal yarns and quick-dry yarns were twisted into twisted yarns using a rotor machine, while the other batch was wrapped into wrapped yarns using a hollow spindle spinning machine. During the process, the twist number was obtained based on the rotating speed difference between the winding and take-up processes. Tables 1–3 show the compositions, parameters, and specifications of wrapped/twisted functional composite yarns. Figure 1 shows the diagrams of a rotor spindle machine and twisted yarn. The bamboo charcoal yarn and the quick-dry yarn were intertwined evenly to form a spiral structure (Figure 1B) using a rotor spin device. Figure 2 shows the assembly of a hollow spindle machine and the structure of a wrapped yarn. A bamboo charcoal yarn and quick-dry yarn were wrapped in two manners, where the bamboo charcoal yarn was the core and the quick-dry yarn was the sheath, and otherwise.

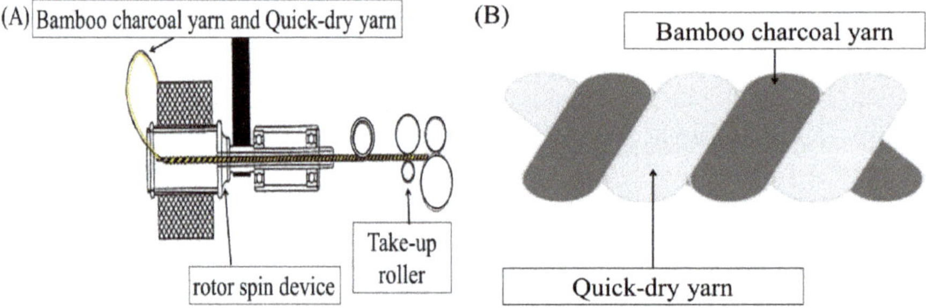

Figure 1. (**A**) The assembly of a rotor spindle machine and (**B**) the structure of a twisted yarns [26].

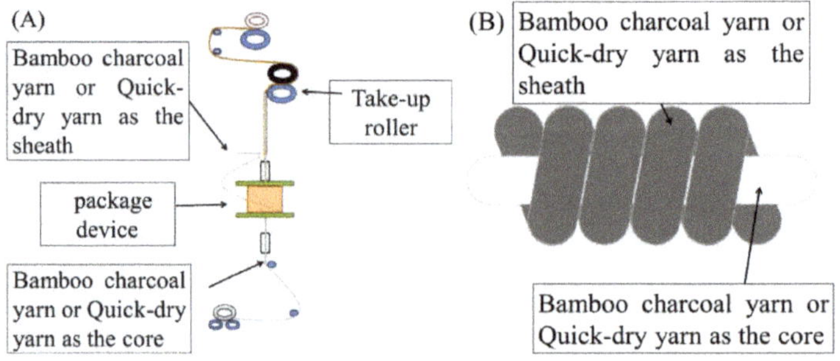

Figure 2. (**A**) assembly of hollow spindle machine and (**B**) diagram of a wrapped yarn [26].

The twist number is computed using the differences in the rotary speed of the package device and take-up roller. The equation is as follows.

$$\text{Twist number} = \frac{R}{T \times D \times \pi} \tag{1}$$

where R is the rotary speed of the rotor spin device or hollow spindle machine (r.p.m.), T is the rotary speed of the take-up roller (r.p.m.), and D is the diameter of the take-up roller (inch).

Table 1. Composition of functional composite yarns. Q: Quick-dry, B: Bamboo charcoal, and D: Denier.

Code	Compositions
Bamboo charcoal yarn	Combination of two 75D bamboo charcoal yarns without twisting or wrapping.
Quick-dry yarn	Combination of two 75D quick-dry yarns without twisting or wrapping.
Q-wrapped yarn	The wrapped yarns have one 75D bamboo charcoal yarn as the core and one 75D quick-dry yarn as the sheath.
B-wrapped yarn	The wrapped yarns have one 75D quick-dry yarn as the core and one 75D bamboo charcoal yarn as the sheath.
B/Q-twisted yarn	The twisted yarns are composed of 75D bamboo charcoal yarn and one 75D quick-dry yarn.

Table 2. Parameters of wrapped/twisted functional composite yarns. Q: Quick-dry, B: Bamboo charcoal, and D: Denier.

Sample Code	Twist Number (T.P.I.)		
Bamboo charcoal yarn		N/A	
Quick-dry yarn		N/A	
B-wrapped yarn	20	25	30
Q-wrapped yarn	20	25	30
B/Q twisted yarn	5	10	15

Note. For twisted yarns, a twist number that was above 15 T.P.I. (twists per inch) caused a high friction between the twisted yarns and the machine, leading to damage to the surface of the twisted yarns. Twist numbers of 20–30 T.P.I. were excluded from the production of twisted yarns. For wrapped yarns, a twist number was below 20 T.P.I., the hollow spindle spinning machine rendered a take-up extension beyond the tolerance of the wrapped yarns. Twist numbers of 5–15 T.P.I. were thus excluded from the production of wrapped yarns.

Table 3. Specification of functional composite yarns. Q: Quick-dry, B: Bamboo charcoal, and D: Denier.

Yarn Type	Twist Number (T.P.I.)	Strength of Yarn (cN)	Linear density (Denier)	Diameter of Yarn (mm)
Bamboo charcoal yarn	0	560.60 ± 30.57	148.90 ± 0.01	0.38 ± 0.01
Quick-dry yarn	0	605.65 ± 28.75	151.50 ± 0.03	0.47 ± 0.04
Q-wrapped yarn	20	610.48 ± 18.36	156.69 ± 0.01	0.28 ± 0.01
	25	555.34 ± 34.70	159.35 ± 0.01	0.26 ± 0.02
	30	542.85 ± 17.42	159.97 ± 0.03	0.21 ± 0.01
B-wrapped yarn	20	635.01 ± 14.82	155.95 ± 0.02	0.23 ± 0.03
	25	616.29 ± 8.16	160.19 ± 0.01	0.21 ± 0.01
	30	599.45 ± 11.95	162.31 ± 0.04	0.17 ± 0.02
B/Q-twisted yarn	5	624.80 ± 16.17	150.77 ± 0.02	0.38 ± 0.03
	10	643.19 ± 14.21	151.51 ± 0.03	0.36 ± 0.04
	15	636.23 ± 14.72	151.98 ± 0.03	0.22 ± 0.02

Note. Standard deviation (SD) is presented in the form of "±".

Functional elastic knits are made with functional composite yarns using a crochet machine. The knits are composed of polyester filament as the warp yarns, rubber thread as the warp inlay yarn, and functional composite yarns as the weft inlay yarn (Figure 3). A total of six types of wrapped yarns and three types of twisted yarns, seen in Table 2. The pure bamboo charcoal yarns and quick-dry yarns serve as the control groups. The fabric specifications are shown in Table 4.

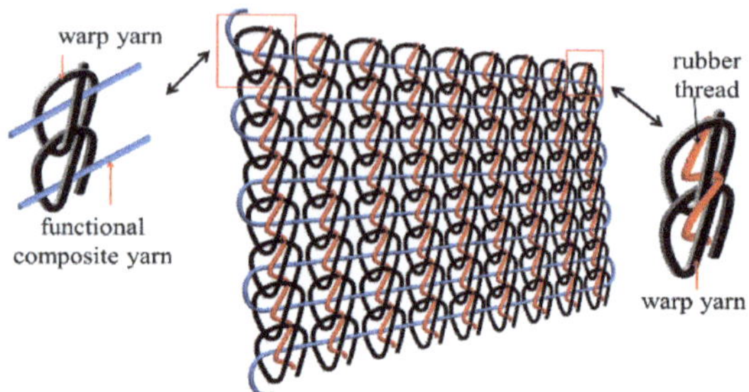

Figure 3. Functional elastic knits where the black yarn is the warp yarn, the red yarn is the rubber thread, and the blue yarn is the functional composite yarns.

Table 4. Specifications of functional elastic knits. Q: Quick-dry, B: Bamboo charcoal, and D: Denier.

Knit Type	Twist Number of the Yarn (T.P.I.)	Weight of the Knits (g/m^2)	Thickness of the Knits (mm)
Bamboo charcoal functional elastic knits	0	409.10	1.06 ± 0.02
Quick-dry functional elastic knits	0	403.70	1.07 ± 0.01
	20	406.30	1.12 ± 0.01
Q-wrapped functional elastic knits	25	411.30	1.13 ± 0.02
	30	411.60	1.12 ± 0.01
	20	405.60	1.14 ± 0.01
B-wrapped functional elastic knits	25	411.00	1.08 ± 0.01
	30	420.40	1.14 ± 0.02
	5	405.90	1.11 ± 0.01
B/Q-twisted functional elastic knits	10	411.70	1.10 ± 0.01
	15	416.00	1.12 ± 0.01

Note. SD is presented in the form of "$\pm$".

2.3. Tests

2.3.1. Mechanical Properties of Yarns

An automatic yarn tester (FPA/M, Statimat-M, Textechno Ltd, Mönchengladbach, Germany) was used to measure the tensile strength and elongation of samples as specified in ASTM D2256. The distance between gauges was 250 mm and tensile rate was 300 mm/min. Twenty samples of each specification are measured in order to have the mean.

2.3.2. Mechanical Properties of Knits

Functional elastic knits were tested for tensile strength using a universal testing machine (HT-2402, Hung Ta Instrument Co., Ltd, Taichung, Taiwan) as specified in ASTM D5034. The distance between gauges was 100 mm and tensile rate was 300 mm/min. Samples had a size of 20 cm × 2.5 cm. Ten samples of each specification were measured in order to have the mean.

2.3.3. Water Vapor Transmission Rate

Functional elastic knits were tested for water vapor transmission rate, following the test standard of ASTM E96 (desiccant method). A piece of sample was mounted on the bottle which has no lid and contained 10 mL of water, and the bottle was placed in the test case for 24 h. The environmental condition of the test box was at a temperature between 23 and 26.7 °C and a relative humidity of 50%. A precision balance was used to measure the weight of the bottle, which was recorded as w_0. After 24-h water evaporation, the bottle was weighed again as w_t. The water vapor transmission rate was computed with the equation as follows.

$$\text{Water Vapor Transmission Rate} = \frac{w_0 - w_t}{A \times t} \times 100\% \tag{2}$$

where w_0 was the initial weight (g) of samples (including the bottle, water, and knits), w_t was the weight (g) of samples after 24-h evaporation (including the bottle, water, and knits), A was the area of knits (m^2), and t was the evaporation duration.

2.3.4. Water Absorption Rate

Functional elastic knits were tested for water absorption rate, following the test standard of AATCC TM 197-2011. Samples had a size of 200 × 25 mm. Five samples were taken respectively along the warp direction and the weft direction. Samples were affixed above the tank with their lower end of 0.5 cm soaking in water for ten minutes. The capillarity of samples along the warp and weft directions was recorded and computed in order to have the mean.

2.3.5. Far-Infrared (FIR) Emissivity

Functional elastic knits were tested for far infrared (FIR) emissivity using a FIR tester (TSS-5X, Desunnano Co., Ltd., Tokyo, Japan). Samples were mounted on the platform of the tester, and the tester was used to measure the FIR emissivity at ten spots. The values were recorded and computed to have the mean. FIR emissivity was presented by the ratios of the absorbed radiant energy to the released radiant energy between samples and the black body. The equation was as follows. The object that totally cannot release radiant energy after it absorbs radiant energy is called the gray body. In contrast, the object that can completely release radiant energy after it absorbs radiant energy is called the black body.

$$\varepsilon = \frac{\text{All radiation energy emitted by a gray body}}{\text{All radiation energy emitted by a black body at the same temperature}} \tag{3}$$

2.3.6. Air Permeability Test

Functional elastic knits are tested for air permeability using an air permeability tester (Textest FX3300, TEXTEST INSTRUMENTS, Zürich, Switzerland), following the test standard of ASTM D737. The size of samples is 25 cm × 25 cm. A sample is fixed behind the air vent and the air permeability against an air pressure of 125 Pa is measured. Twelve samples of each specification are used for the measurement, and the values are recorded and computed in order to have the mean.

2.3.7. Bursting Strength

Functional elastic knit were tested for bursting strength using a universal testing machine (HT-2402, Hung Ta Instrument Co., Ltd., Taichung, Taiwan), following the test standard of ASTM D3787. Samples were affixed in a ring clamp, and then tested using a pulling clamp at speed of 305 ± 13 mm/min. The samples had a size of 125 mm × 125 mm. Five samples for each specification were tested and the mean was recorded.

3. Results and Discussion

3.1. Effects of Twist Number on the Mechanical Properties of Yarns

Figure 4 shows that the twisting process provides functional composite yarns with greater tenacity, as compared to the tenacity of the control group (i.e., pure bamboo charcoal yarns and quick-dry yarns). Conversely, the wrapping process provided functional composite yarns with greater tenacity, as compared to the tenacity of the control group. Bamboo charcoal yarns had a low tenacity and elongation at the break, according to Table 3 and Figure 5. The wrapping and twisting process improved the extensibility of the yarns. Therefore, regardless of whether it was wrapped yarns or twisted yarns, the elongation at the break of the functional composite yarns was proportional to the twist number of the yarns.

Moreover, the tenacity of functional composite yarns that were processed with a wrapping process showed a remarkable decrease when the twist number increases. For wrapped yarns, the wrap material could only enwrap the core, but it was primarily the core that bears the majority of the exerted force. Therefore, variations in twist number did not have a considerable influence on the tenacity of the wrapped yarns [26]. Figures 1B and 2B are the structural diagrams of twisted yarns and wrapped yarns. The wrapped yarns were composed of bamboo charcoal yarn and quick-dry yarn alternatively as the core and sheath. A low twist number was unable to enwrap the core completely. After the breakage of the core, a large amount of sheath was possible to slip. A high twist number compactly enwrapped the core, decreasing the slip level of sheath. The twisted yarns were composed of bamboo charcoal yarns and quick-dry yarns with a spiral structure. The higher the twist number, the denser the spiral structure. When a tensile force was exerted, the twisted yarns exhibited a structural slip where the spiral structures were presented. As a result, increasing the twist number improved the elongation of the twisted yarns. However, a high twist number also increased the contact area between the two yarns, as was the case with the friction. Therefore, a high twist number inhibited the slip of structure of twisted yarns, which eventually caused a lower elongation at the break, more than that of the wrapped yarns.

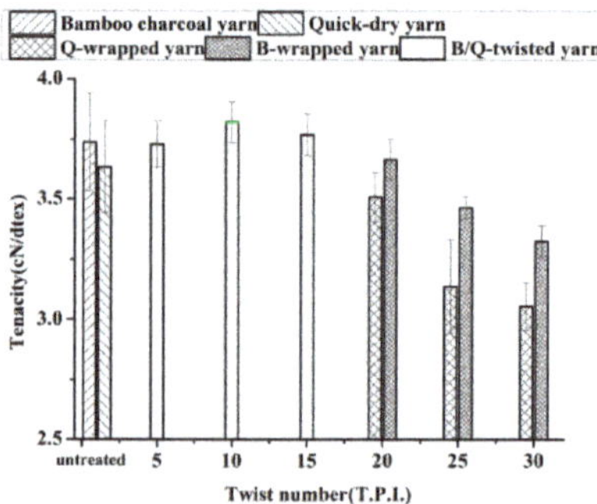

Figure 4. Tenacity of functional composite yarns in relation to the administration of a wrapping/twisting process as well as the twist number. The error bar indicates the Standard deviation (SD). Q: Quick-dry, and B: Bamboo charcoal.

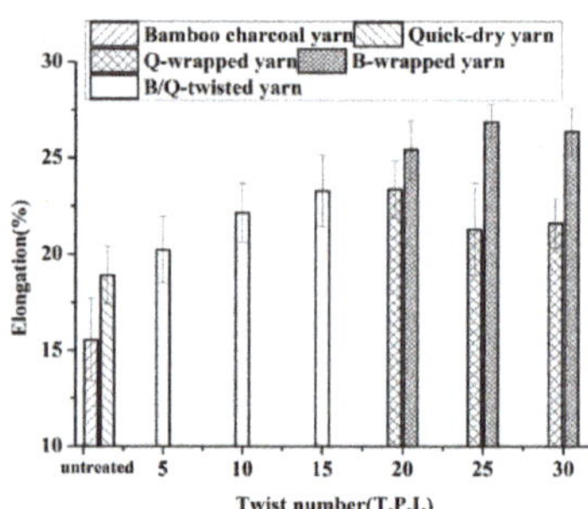

Figure 5. Elongation of functional composite yarns in relation to the administration of a wrapping/twisting process as well as the twist number. The error bar indicates the SD.

3.2. Effects of Twist Number on the Mechanical Properties of Functional Elastic Knits

Functional composite yarns were weft inlaid into functional elastic knits using a crochet machine. Therefore, mechanical properties of knits were correlated with the parameters of the functional composite yarns. In Figure 6a, there are no noticeable differences in the tensile stress along the warp direction of the elastic knits. Figure 6b shows that there are no critical differences in the tensile stress along the weft direction of the elastic knits when they are composed of B/Q twisted yarns at a different twist number.

This result was ascribed to the peculiar structure of the knits, which prevented the weft inlay yarn from interfering with the mechanical properties along the warp direction. This was also the case with the elongation along the warp direction as seen in Figure 7a. All functional elastic knits had similar elongation along the warp direction. When composed of a twist number between 5 and 15 T.P.I., B/Q-twisted yarns had similar tenacity, which subsequently provided the elastic knits with similar tensile stress. In contrast, for the elastic knits that were composed of wrapped yarns, the tensile stress along the weft direction was largely in an inverse proportion to the twist number of the wrapped yarns. A high twist number rendered wrapped yarns with lower mechanical properties, which thus had a negative influence on the elastic knits. Moreover, the elongation along the weft direction of elastic knits was dependent on the elongation of the wrapped yarns. Compared to the twisted yarns, the wrapped yarns provided the yarns with higher elongation as seen in Figure 5. Therefore, elastic knits that were made of wrapped yarns had a greater elongation along the weft direction than the elastic knits that were made of twisted yarns (Figure 7b).

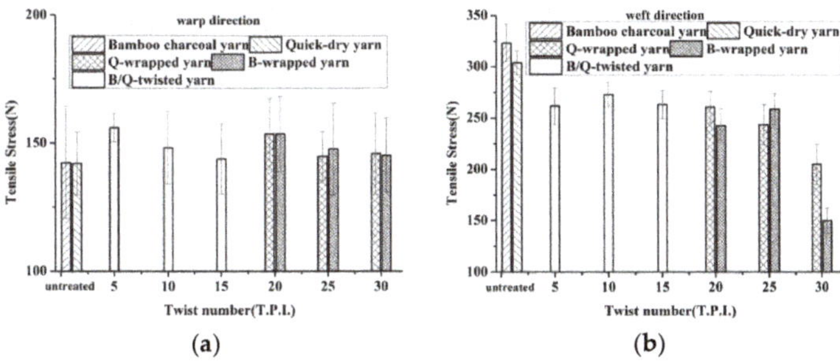

Figure 6. Tensile stress along (**a**) the warp direction and (**b**) weft direction of functional elastic knits in relation to the administration of a wrapping/twisting process, as well as the twist number of the yarns. The error bar indicates the SD.

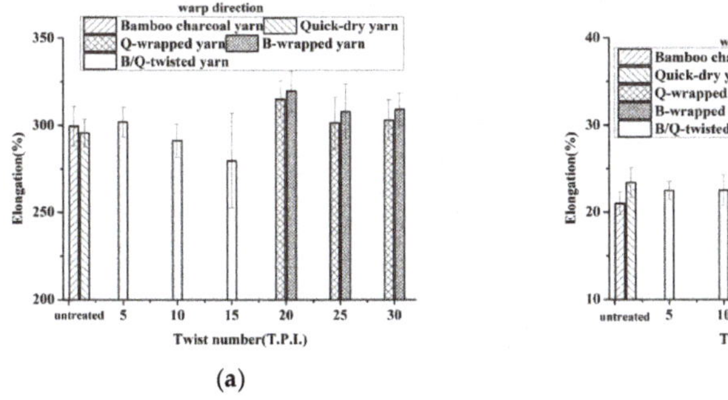

Figure 7. Elongation along (**a**) the warp direction and (**b**) weft direction of functional elastic knits in relation to the administration of a wrapping/twisting process as well as the twist number of the yarns. The error bar indicates the SD.

3.3. Effects of Twist Number on the Bursting Strength of Functional Elastic Knits

Figure 8 shows that both the experimental and control groups have a bursting strength between 170 and 180 N, while Figure 9 shows that samples that undergo the bursting test all exhibit failure along the weft direction, rather than the warp direction.

Because the functional elastic knits were composed of rubber threads along the warp direction, this provided the elastic knits with good extension along the warp direction. As a result, the rubber threads of elastic knits transmitted the externally applied force via local deformations. On the other hand, the functional elastic knits were composed of weft yarns in order to form an inlay structure as seen in blue lines in Figure 3, and the knits failed to extend along the weft direction, along which the prior damage was rendered. In addition, the functional elastic knits had comparable bursting stress regardless of using yarns for different specifications. This result resembled that which was described in Section 3.2. It was primarily due to the similar mechanical properties of the different functional yarns, as seen in Figures 4 and 5. Therefore, with a specified weft density of 64 pick/inch, the functional yarn types did not have a distinctive difference in the bursting stress of the elastic knits.

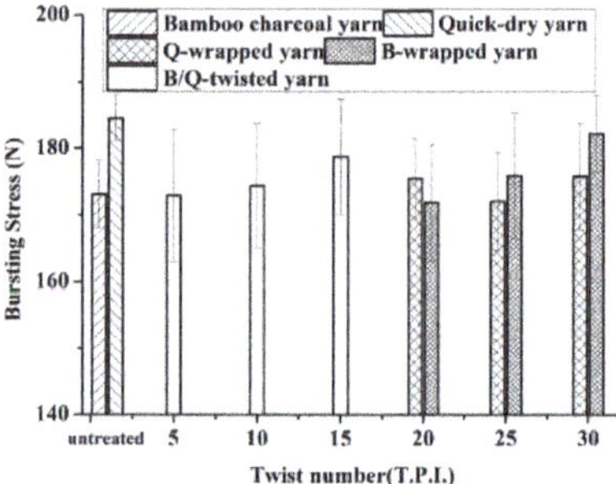

Figure 8. Bursting stress of elastic knits as related to parameters of different functional yarns. The error bar indicates the SD.

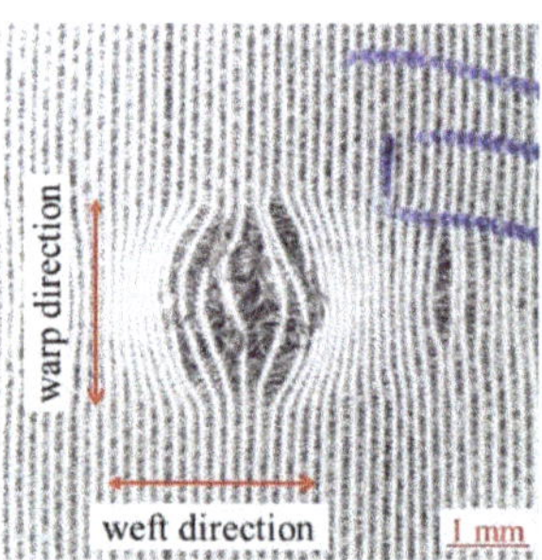

Figure 9. Images of the fractured elastic knits after the bursting strength test. As all samples exhibit similar failure over the surface, B/Q-twisted yarns made of 20 T.P.I. is used in this diagram.

3.4. Effects of Twist Number on the Air Permeability of Functional Elastic Knits

Doing exercise accelerates metabolism and demands 20% of the total energy consumed by metabolism. At the same time, the rest 80% is dissipated via the form of heat. If the human body fails to dissipate heat in a short time, the core temperature of the body then increases by 5 °C every ten minutes. When the core temperature reaches to a certain point, the human body perspires in order to attain a thermal balance [11]. This phenomenon leads to dehydration, and the air permeability of the elastic knits is thus a crucial factor in terms of the comfort on the skin [27,28].

In comparison to the control groups (i.e., functional elastic knits that were composed of non-processed bamboo charcoal yarns or non-processed quick-dry yarns, the elastic knits that were composed of twisted yarns or wrapped yarns had a considerably higher air permeability. In addition, the analysis software of an optical microscope showed that the control groups had an average pore size of 0.03–0.04 mm^2, while the experimental groups (i.e., functional elastic knits) had an average pore size 0.08–0.1 mm^2. There were no distinctive variations in the thickness and weight of the knits when the diameter of the yarns decreases. The results were ascribed to the fact that the thickness and weight of the elastic knits were primarily based on the amount of rubber threads. Therefore, using different wrapped yarns or twisted yarns did not noticeably change the weight and thickness of the elastic knits.

Regardless of whether it was bamboo charcoal yarns or quick-dry yarns, they had a fluffy structure. When they were used to produce elastic knits using a crochet machine, the knits demonstrated a shrinkage when they were off the pre-tension force exerted by the machine. Therefore, the pores of the elastic knits were compressed as seen in Figure 10A,B. The wrapping or twisting process made the structure of yarns compact, resulting in a smaller diameter. Therefore, the resulting elastic knits had a larger pore size as seen in Figure 10C–K. Previous studies indicated that the variations in the diameter of yarn as well as variations in the thickness and weight of the knits affected the air permeability of the knits. In comparison of Table 2 and Figure 11, the air permeability of the elastic knits was inversely proportional to the diameter of yarns. This result was in conformity with that of previous studies [12,29].

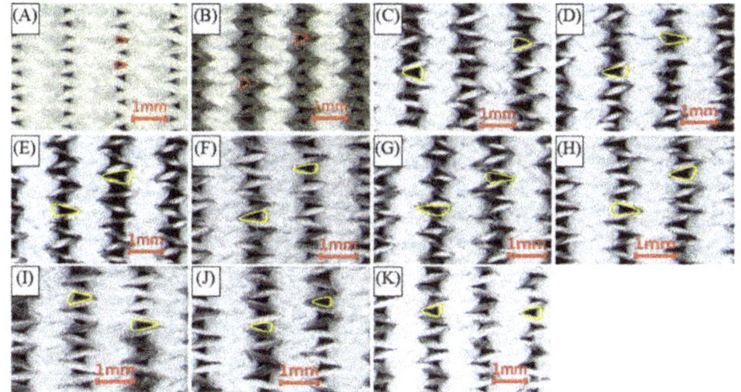

Figure 10. Surficial observation of the functional elastic knits that are made of (**A**) quick-dry yarn; (**B**) bamboo charcoal yarn; (**C**) 20 T.P.I. B-wrapped yarn; (**D**) 25 T.P.I. B-wrapped yarn; (**E**) 30 T.P.I. B-wrapped yarn; (**F**) 20 T.P.I. Q-wrapped yarn; (**G**) 25 T.P.I. Q-wrapped yarn; (**H**) 30 T.P.I. Q-wrapped yarn; (**I**) 5 T.P.I. B/Q-twisted yarn; (**J**) 10 T.P.I. B/Q-twisted yarn; and (**K**) 15 T.P.I. B/Q-twisted yarn. The small pore size is indicated in red drop-like triangles and the large pore size is indicated yellow drop-like triangles.

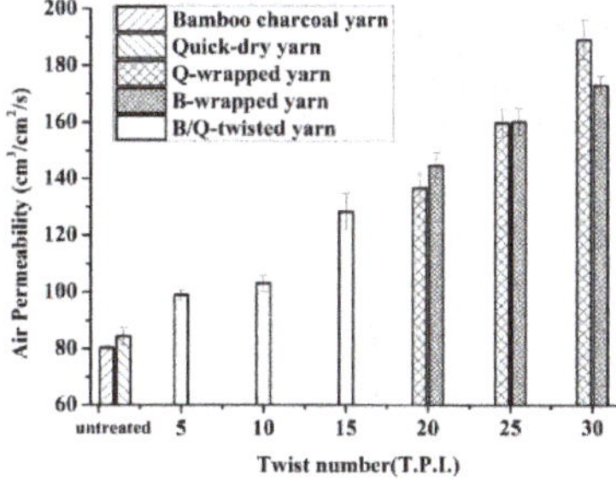

Figure 11. Air permeability of functional elastic knits in relation to the administration of a wrapping/twisting process as well as the twist number of the yarns. The error bar indicates the SD.

3.5. Effects of Twist Number on the Water Vapor Transmission Rate of Functional Elastic Knits

Water vapor adsorbs to the surface of knits when they come into contact. The knits diffuse and evaporate the moisture. If water vapor fails to penetrate the knits, it undergoes condensation in the air layer between the skin and the knit. The residual moisture makes the wearers uncomfortable when their body temperature decreases [11,28]. The skin of the human body has a water vapor transmission rate of 215–350 g/m^2 day [11,30–32], and any lower water vapor transmission rates of knits would retain moisture. The water vapor transmission rate of knits was dependent on the types of yarns, specification of the yarns, and the structure and thickness of the knits [12,32]. Hence, this study investigated the effects of two parameters, namely, the type of yarns and the specification of the yarns.

Figure 12 shows that all functional elastic knits have a high water vapor transmission rate. The control groups had a greater water vapor transmission rate than the experimental groups. Moreover, the control group of non-processed bamboo charcoal yarns outperformed the control group of non-processed quick-dry yarns, particularly in terms of the water vapor transmission rate. For the experimental groups, the water vapor transmission rate of the elastic knits had a decreasing trend when their constituent yarns were composed of a high twist number. The functional composite yarns had a firm structure after they were twisted or wrapped. However, the advantages of bamboo charcoal yarns and quick-dry yarns were no longer present when they were firmly compressed at a twist number beyond 15 T.P.I., which eventually caused a similar water vapor transmission rate of functional elastic knits.

The adsorption and evaporation of the moisture were dependent on the micro-structure and the hydrophilic properties of different yarns. Non-processed bamboo charcoal yarns had a porous microstructure that helped with the seizure of vapors. Similarly, non-processed quick-dry yarns had a channeled-like structure over the surface, which increased the specific areas. Therefore, both fibers had good diffusion and evaporation. In sum, for the experimental groups, regardless of whether it was the wrapped or the twisted yarns, all the functional elastic knits attained a water vapor transmission rate that was qualified for use on human skin, preventing condensation of the water vapor.

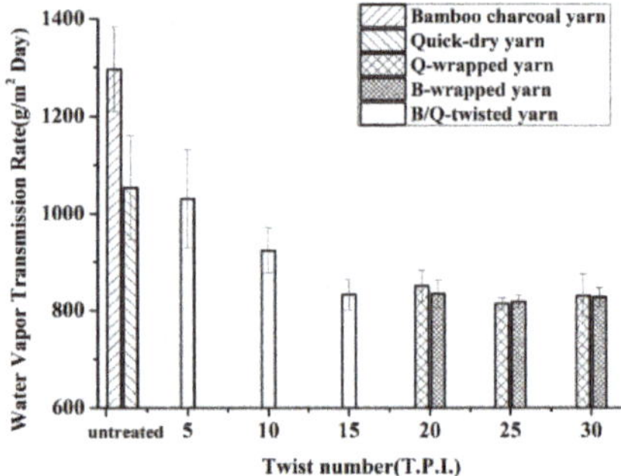

Figure 12. Water vapor transmission rate of functional elastic knits in relation to the administration of a wrapping/twisting process as well as the twist number of the yarns. The error bar indicates the SD.

3.6. Effects of Twist Number on the Water Absorption Rate of Functional Elastic Knits

In order for the comfort of wearers, the elastic knits should have both efficient water absorption and evaporation. Therefore, two functional yarns with a micro-structure were used in this study and their influence on the water absorption rate of the functional elastic knits was examined.

Figure 13 shows that the water absorption rate along the warp direction is noticeably different from that along the weft direction. The functional composite yarns were weft inlaid into functional elastic knits, which resulted in a distinctive increase in water absorption along the weft direction. Figure 13b shows that the elastic knits of Q-wrapped yarns at 20 T.P.I. have water absorption along the weft direction of 80–100 mm, while all the other elastic knits have water absorption along the weft direction of 100–130 mm. Nevertheless, Figure 13a shows that all the elastic knits that have water absorption along the warp direction is lower than 50 mm.

These results were ascribed to the fabrication structure of the elastic knits that was composed of the weft-inlaid functional composite yarns and warp yarn of PET yarns. Functional composite yarns had a high capillary phenomenon, which contributed to water absorption along the weft direction. In contrast, the PET yarns had low water absorption, which resulted in a discontinuity of the capillary phenomenon along the warp direction, and a low water absorption along the warp direction.

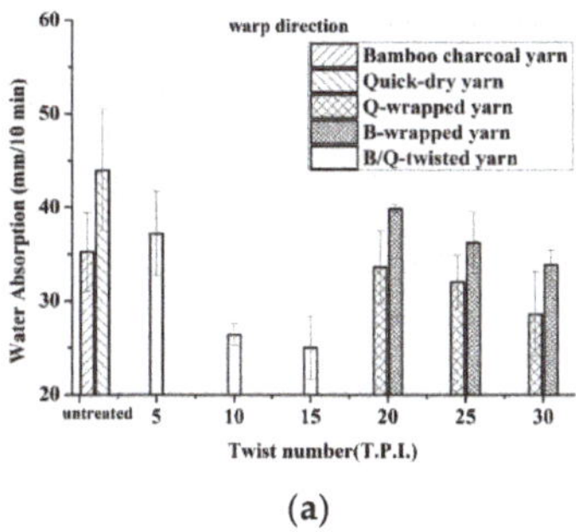

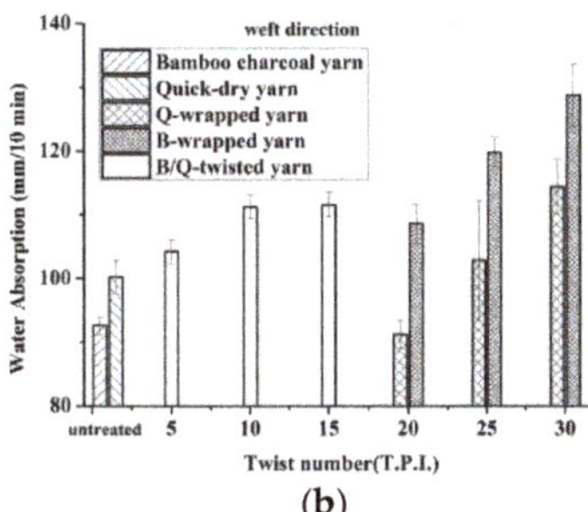

Figure 13. Water absorption along (**a**) the warp direction and (**b**) weft direction of functional elastic knits in relation to the administration of a wrapping/twisting process as well as twist number of the yarns. The error bar indicates the SD.

3.7. Effects of Twist Number on the Far-Infrared Emissivity of Functional Elastic Knits

FIR rays are invisible lights between 5.6 and 1000 μm. The human body absorbs FIR rays that help with molecular rotation and vibrational effects. The blood microcirculation of the irradiated regions also causes vessels to expand and increase the temperature of the body [10]. Figure 14 shows that, in addition to the elastic knits that are composed of pure quick-dry yarns, other functional elastic knits that are composed of bamboo charcoal yarns all have FIR emissivity of above 0.8. According to the test standard of FTTS-FA-010, fabrics with an FIR emissivity above 0.8 are qualified for the application in FIR emissivity.

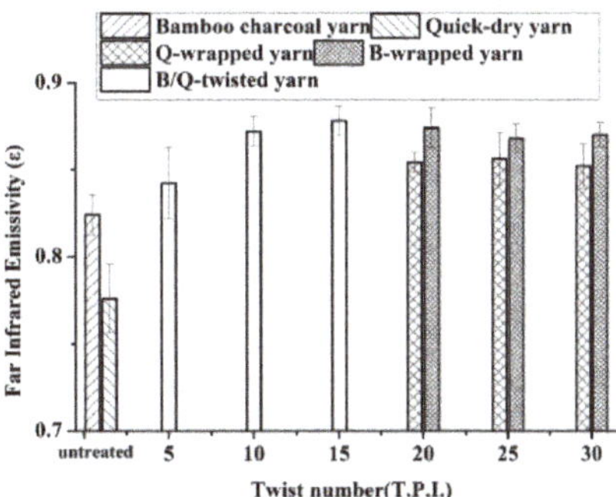

Figure 14. Far infrared ray (FIR) emissivity of functional elastic knits in relation to the administration of a wrapping/twisting process, as well as the twist number of the yarns. The error bar indicates the SD.

In comparison of the control group (i.e., elastic knits that were composed of bamboo charcoal yarn), the experimental groups (i.e., functional elastic knits that were composed of Q-wrapped yarns,

B-wrapped yarns, and B/Q-twisted yarns) had greater FIR emissivity. In addition, the functional elastic knits had a higher FIR emissivity when they were composed of B-wrapped yarns, as opposed to Q-wrapped yarns. B-wrapped yarns had bamboo charcoal yarns as the sheath and quick-dry yarns as the core, while B-wrapped yarns were made the other way around.

This result was due to the fact that the wrapping or twisting process increased the content of the bamboo charcoal yarn in the functional composite knits, thereby improving the FIR emissivity. The Q-wrapped yarns were composed of a smaller amount of bamboo charcoal yarn, and the wrapping process further decreased their diameter, debilitating the radiation of FIR rays. Moreover, a previous study indicated that increasing the numbers of lamination layers increased the contact area with FIR radiation, and thus, improved the FIR emissivity [33].

4. Conclusions

This study investigates the functional composite yarns and functional elastic knits in terms of physical properties and functions. The effects of the twist number of the functional composite yarns were examined in order to increase the surface of functional yarns per unit area and then improve the mechanical properties. The test results showed that using a wrapping or a twisting process provided functional composite yarns with good mechanical properties, and that the former outperformed the latter in terms of the elongation of the functional composite yarns.

In addition, the properties of functional elastic knits were evaluated in order to examine the influence of the parameters of the yarns. The test results showed that using twisted yarns or wrapped yarns resulted in a decrease in water vapor transmission rate of the functional elastic knits but an increase in their FIR emissivity, water absorption rate, and air permeability. According to production cost, mechanical properties, water vapor transmission rate, air permeability, water absorption, and FIR emissivity, the optimal parameter of functional elastic knits was B-wrapped yarns at 20 T.P.I. The functional elastic knits thus had adjustable fabric shrinkage, which helped the wearers to support joints and muscles during exercises.

Acknowledgments: The authors would especially like to thank the Ministry of Science and Technology of Taiwan, for financially supporting this research under Contract MOST 105-2622-E-166-001-CC2.

Author Contributions: In this study, the concepts and designs for the experiment, all required materials, as well as processing and assessment instruments, were provided by Jia-Horng Lin and Ching-Wen Lou. Data were analyzed, and experimental results were examined by Chih Hung He and Yu-Tien Huang. The experiment was conducted and the text was composed by Chih Hung He.

Conflicts of Interest: The authors declare no conflict of interest.

References

1. Gambichler, T. Ultraviolet protection of clothing. In *Functional Textiles for Improved Performance, Protection and Health*; Woodhead Publishing: Cambridge, UK, 2011; pp. 45–63.
2. Zhang, X. Antistatic and conductive textiles. In *Functional Textiles for Improved Performance, Protection and Health*; Woodhead Publishing: Cambridge, UK, 2011; pp. 27–44.
3. Kothari, V.K.; Bhattacharjee, D. Artificial neural network modelling for prediction of thermal transmission properties of woven fabrics. In *Soft Computing in Textile Engineering*; Woodhead Publishing: Cambridge, UK, 2011; pp. 403–423.
4. Lawrence, C. Chapter 10—Fibre to yarn: Filament yarn spinning. In *Textiles and Fashion*; Woodhead Publishing: Cambridge, UK, 2015; pp. 213–253.
5. Li, T.T.; Pan, Y.J.; Hsieh, C.T.; Lou, C.W.; Chuang, Y.C.; Huang, Y.T.; Lin, J.H. Comfort and functional properties of far-infrared/anion-releasing warp-knitted elastic composite fabrics using bamboo charcoal, copper, and phase change materials. *Appl. Sci.* **2016**, *6*, 62. [CrossRef]
6. Dong, Y.; Kong, J.; Mu, C.; Zhao, C.; Thomas, N.L.; Lu, X. Materials design towards sport textiles with low-friction and moisture-wicking dual functions. *Mater Des.* **2015**, *88*, 82–87. [CrossRef]

7. Li, Y. Perceptions of temperature, moisture and comfort in clothing during environmental transients. *Ergonomics* **2005**, *48*, 234–248. [CrossRef] [PubMed]

8. Classen, E. Comfort testing of textiles. In *Advanced Characterization and Testing of Textiles*; Vermeersch, O., Izquierdo, V., Eds.; Woodhead Publishing: Cambridge, UK, 2018; pp. 59–69.

9. McCann, J. Environmentally conscious fabric selection in sportswear design. In *Textiles for Sportswear*; Woodhead Publishing: Cambridge, UK, 2015; pp. 17–52.

10. Dyer, J. Infrared functional textiles. In *Functional Textiles for Improved Performance, Protection and Health*; Woodhead Publishing: Cambridge, UK, 2011; pp. 184–197.

11. Daanen, H. Physiological strain and comfort in sports clothing. In *Textiles for Sportswear*; Woodhead Publishing: Cambridge, UK, 2015; pp. 153–168.

12. Majumdar, A.; Mukhopadhyay, S.; Yadav, R. Thermal properties of knitted fabrics made from cotton and regenerated bamboo cellulosic fibres. *Int. J. Therm. Sci.* **2010**, *49*, 2042–2048. [CrossRef]

13. Deopura, B.L.; Padaki, N.V. Chapter 5—Synthetic textile fibres: Polyamide, polyester and aramid fibres a2 -sinclair, rose. In *Textiles and Fashion*; Woodhead Publishing: Cambridge, UK, 2015; pp. 97–114.

14. Alexander, J.; Selfe, J.; Oliver, B.; Mee, D.; Carter, A.; Scott, M.; Richards, J.; May, K. An exploratory study into the effects of a 20 minute crushed ice application on knee joint position sense during a small knee bend. *Phys. Ther. Sport* **2016**, *18*, 21–26. [CrossRef] [PubMed]

15. Kobara, K.; Fujita, D.; Osaka, H.; Ito, T.; Watanabe, S. Influence of distance between the rotation axis of back support and the hip joint on shear force applied to buttocks in a reclining wheelchair's back support. *Prosthet. Orthot. Int.* **2013**, *37*, 459–464. [CrossRef] [PubMed]

16. Jorgensen, M.J.; Marras, W.S. The effect of lumbar back support tension on trunk muscle activity. *Clin. Biomech.* **2000**, *15*, 292–294. [CrossRef]

17. Rønning, R.; Rønning, I.; Gerner, T.; Engebretsen, L. The efficacy of wrist protectors in preventing snowboarding injuries. *Am. J. Sports Med.* **2001**, *29*, 581–585. [CrossRef] [PubMed]

18. Stuart, P.; Briggs, P. Closed extensor tendon rupture and distal radial fracture with use of a gymnast's wrist support. *Br. J. Sports Med.* **1993**, *27*, 92–93. [CrossRef] [PubMed]

19. Lin, C.M.; Lin, C.W.; Yang, Y.C.; Lou, C.W.; Chen, A.P.; Lin, J.H. Evaluation of the manufacturing and functions of complex yarn and fabrics. *Fibres Text. East. Eur.* **2012**, *20*, 47–50.

20. Lin, J.H.; Lin, C.M.; Huang, C.H.; Chen, A.P.; Chen, C.P.; Lou, C.W. Physical properties of electrically conductive complex-ply yarns and woven fabrics made from recycled polypropylene. *J. Eng. Fibers Fabr.* **2013**, *8*, 30–38.

21. Lin, J.H.; Huang, Y.T.; Li, T.T.; Lin, C.M.; Lou, C.W. Bamboo charcoal/phase change material/stainless steel ring-spun complex yarn and its far-infrared/anion-releasing elastic warp-knitted fabric: Fabrication and functional evaluation. *J. Ind. Text.* **2016**, *46*, 624–642. [CrossRef]

22. Yu, Z.C.; Zhang, J.F.; Lou, C.W.; Lin, J.H. Wicking behavior and antibacterial properties of multifunctional knitted fabrics made from metal commingled yarns. *J. Text. Inst.* **2015**, *106*, 862–871. [CrossRef]

23. Lou, C.W.; Hu, J.J.; Lu, P.C.; Lin, J.H. Effect of twist coefficient and thermal treatment temperature on elasticity and tensile strength of wrapped yarns. *Text. Res. J.* **2016**, *86*, 24–33. [CrossRef]

24. Liu, T.; Choi, K.F.; Li, Y. Wicking in twisted yarns. *J. Colloid Interface Sci.* **2008**, *318*, 134–139. [CrossRef] [PubMed]

25. Perwuelz, A.; Casetta, M.; Caze, C. Liquid organisation during capillary rise in yarns—Influence of yarn torsion. *Polym. Test.* **2001**, *20*, 553–561. [CrossRef]

26. Qu, H.; Skorobogatiy, M. 2—Conductive polymer yarns for electronic textiles. In *Electronic textiles*; Woodhead Publishing: Oxford, UK, 2015; pp. 21–53.

27. Tessier, D. Testing thermal properties of textiles a2-dolez, patricia. In *Advanced Characterization and Testing of Textiles*; Vermeersch, O., Izquierdo, V., Eds.; Woodhead Publishing: Cambridge, UK, 2018; pp. 71–92.

28. Fung, W. Coated and laminated textiles in sportswear. In *Textiles in Sport*; Woodhead Publishing: Cambridge, UK, 2005; pp. 134–174.

29. Van Amber, R.R.; Wilson, C.A.; Laing, R.M.; Lowe, B.J.; Niven, B.E. Thermal and moisture transfer properties of sock fabrics differing in fiber type, yarn, and fabric structure. *Text. Res. J.* **2015**, *85*, 1269–1280. [CrossRef]

30. Lamke, L.O.; Nilsson, G.E.; Reithner, H.L. The evaporative water loss from burns and the water-vapour permeability of grafts and artificial membranes used in the treatment of burns. *Burns* **1977**, *3*, 159–165. [CrossRef]

31. RuizCardona, L.; Sanzgiri, Y.D.; Benedetti, L.M.; Stella, V.J.; Topp, E.M. Application of benzyl hyaluronate membranes as potential wound dressings: Evaluation of water vapour and gas permeabilities. *Biomaterials* **1996**, *17*, 1639–1643. [CrossRef]
32. Li, Y. The science of clothing comfort. *Text. Prog.* **2001**, *31*, 1–135. [CrossRef]
33. Lin, J.H.; Chen, A.P.; Hsieh, C.T.; Lin, C.W.; Lin, C.M.; Lou, C.W. Physical properties of the functional bamboo charcoal/stainless steel core-sheath yarns and knitted fabrics. *Text. Res. J.* **2011**, *81*, 567–573. [CrossRef]

MDPI

St. Alban-Anlage 66

4052 Basel

Switzerland

Tel. +41 61 683 77 34

Fax +41 61 302 89 18

www.mdpi.com

Applied Sciences Editorial Office

E-mail: applsci@mdpi.com

www.mdpi.com/journal/applsci